Electrical Science 3
Checkbook

J O Bird
BSc(Hons), AFIMA, TEng(CEI), MITE

A J C May
BA, CEng, MIMechE, FITE, MBIM

J R Penketh
BSc, CEng, MIEE

Butterworths
London Boston Sydney Wellington Durban Toronto

First published 1981

© Butterworth & Co (Publishers) Ltd, 1981

British Library Cataloguing in Publication Dada

Bird, J. O.
    Electrical science 3 checkbook.
    1. Electric engineering
    I. Title   II. May, A. J. C.
    III. Penketh, J. R.
    621.3'03   TK145

    ISBN 0-408-00657-9
    ISBN 0-408-00626-9 Pbk

621.3

Typeset by Scribe Design, Gillingham, Kent
Printed in Scotland by Thomson Litho Ltd, East Kilbride

Contents

Note to Reader

As textbooks become more expensive, authors are often asked to reduce the number of worked and unworked problems, examples and case studies. This may reduce costs, but it can be at the expense of practical work which gives point to the theory.

Checkbooks if anything lean the other way. They let problem-solving establish and exemplify the theory contained in technician syllabuses. The Checkbook reader can gain *real* understanding through seeing problems solved and through solving problems himself.

Checkbooks do not supplant fuller textbooks, but rather supplement them with an alternative emphasis and an ample provision of worked and unworked problems. The brief outline of essential data—definitions, formulae, laws, regulations, codes of practice, standards, conventions, procedures, etc—will be a useful introduction to a course and a valuable aid to revision. Short-answer and multi-choice problems are a valuable feature of many Checkbooks, together with conventional problems and answers.

Checkbook authors are carefully selected. Most are experienced and successful technical writers; all are experts in their own subjects; but a more important qualification still is their ability to demonstrate and teach the solution of problems in their particular branch of technology, mathematics or science.

Authors, General Editors and Publishers are partners in this major low-priced series whose essence is captured by the Checkbook symbol of a question or problem 'checked' by a tick for correct solution.

Note to Reader

Preface

This textbook of worked problems provides coverage of the Technician Education Council's level 3 unit in Electrical Science (syllabus U75/062). However it can also be regarded as a basic textbook in Electrical Science for a much wider range of courses.

The aim of the book is to increase the mechanical engineering technician's knowledge of applications in industry of electrical and electronic systems. Each topic considered in the text is presented in a way that assumes in the reader only the electrical knowledge attained at TEC level 2 in Engineering Science (U75/053).

This practical electrical science book contains over 100 illustrations and some 100 detailed worked problems, followed by nearly 400 further problems with answers.

The authors would like to express their appreciation for the friendly co-operation and helpful advice given to them by the publishers. Thanks are also due to Mrs Elaine Mayo for the excellent typing of the manuscript.

Finally, the authors would like to add a special word of thanks to their wives, Elizabeth, Juliet and Marilyn for their patience, help and encouragement during the preparation of this book.

JO Bird
AJC May
JR Penketh
Highbury College of Technology
Portsmouth

Butterworths Technical and Scientific Checkbooks

General Editors for Science, Engineering and Mathematics titles:
J.O. Bird and A.J.C. May, Highbury College of Technology, Portsmouth.

General Editor for Building, Civil Engineering, Surveying and Architectural titles:
Colin R. Bassett, lately of Guildford County College of Technology.

A comprehensive range of Checkbooks will be available to cover the major syllabus areas of the TEC, SCOTEC and similar examining authorities. A comprehensive list is given below and classified according to levels.

Level 1 (Red covers)
Mathematics
Physical Science
Physics
Construction Drawing
Construction Technology
Microelectronic Systems
Engineering Drawing
Workshop Processes & Materials

Level 2 (Blue covers)
Mathematics
Chemistry
Physics
Building Science and Materials
Construction Technology
Electrical & Electronic Applications
Electrical & Electronic Principles
Electronics
Microelectronic Systems
Engineering Drawing
Engineering Science
Manufacturing Technology

Level 3 (Yellow covers)
Mathematics
Chemistry
Building Measurement
Construction Technology
Environmental Science
Electrical Principles
Electronics
Electrical Science
Mechanical Science
Engineering Mathematics & Science

Level 4 (Green covers)
Mathematics
Building Law
Building Services & Equipment
Construction Site Studies
Concrete Technology
Economics of the Construction Industry
Geotechnics
Engineering Instrumentation & Control

Level 5
Building Services & Equipment
Construction Technology

1 Machines

A. MAIN POINTS CONCERNED WITH D.C. MACHINES

1   A machine is a system having an input and an output.
2   (i) When the input to an electrical machine is electrical energy (seen as applying a voltage to the electrical terminals of the machine) and the output is mechanical energy (seen as a rotating shaft), the machine is called an electric **motor**. Thus an electric motor converts electrical energy into mechanical energy.

(ii) When the input to an electrical machine is mechanical energy (seen as, say, a diesel motor coupled to the machine by a shaft) and the output is electrical energy (seen as a voltage appearing at the electrical terminals of the machine), the machine is called a **generator**. Thus, a generator converts mechanical energy to electrical energy. It can be seen that a motor and a generator are the same type of system, but with inputs and outputs reversed.

3   The efficiency of an electrical machine is the ratio of the output power to the input power and is usually expressed as a percentage. The Greek letter, *eta*, η is used to signify efficiency and as the units are power/power, then efficiency has no units. Thus

Efficiency, $\eta = \dfrac{\text{output power}}{\text{input power}} \times 100$ **per cent**

4   **The action of a commutator** In an electric motor, conductors rotate in a uniform magnetic field. A single-loop conductor mounted between permanent magnets is shown in *Fig 1*. A voltage is applied at points A and B in *Fig 1 (a)*.

A force, F, acts on the loop due to the interaction of the magnetic field of the permanent magnets and the magnetic field created by the current flowing in the loop. This force is proportional to the flux density, B, the current flowing, I, and the effective length of the conductor, l, i.e., $F = BIl$. The force is made up of two parts, one acting vertically downwards due to the current flowing from C to D and the other acting vertically upwards due to the current flowing from E to F (from Fleming's left-hand rule). If the loop is free to rotate, then when it has rotated through 180°, the conductors are as shown in *Fig 1(b)*. For rotation to continue in the same direction, it is necessary for the current flow to be as shown in *Fig 1(b)*, i.e. from D to C and from F to E. This apparent reversal in the direction of current flow is achieved by a process called **commutation**. With reference to *Fig 2(a)*, when a direct voltage is applied at A and B, then as the single-loop conductor rotates, current flow will always be away from the commutator for the part of the conductor adjacent to the N-pole and towards the commutator for

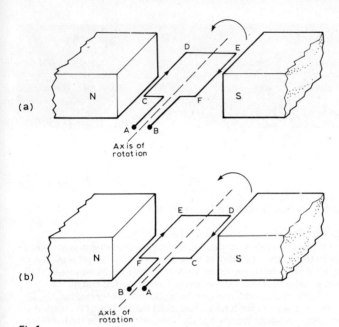

(a)

(b)

Axis of
rotation

Axis of
rotation

Fig 1

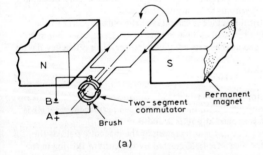

Permanent
magnet

Two-segment
commutator

B
A +

Brush

(a)

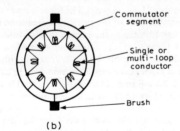

Commutator
segment

Single or
multi-loop
conductor

Brush

Fig 2   (b)

2

the part of the conductor adjacent to the S-pole. Thus the forces act to give continuous rotation in an anti-clockwise direction. The arrangement shown in *Fig 2* is called a 'two-segment' commutator and the voltage is applied to the rotating segments by stationary **brushes** (usually carbon blocks), which slide on the commutator material (usually copper), when rotation takes place.

In practice, there are many conductors on the rotating part of a d.c. machine and these are attached to many commutator segments. A schematic diagram of a multi-segment commutator is shown in *Fig 2(b)*.

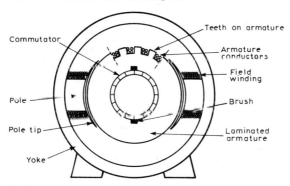

Fig 3

5 **D.C. machine construction** The basic parts of any d.c. machine are shown in *Fig 3*, and comprise:

  (a) a stationary part called the **stator**, having:

      (i) a steel ring called the **yoke**, to which are attached

     (ii) the magnetic **poles**, around which are the

    (iii) **field windings**, i.e. many turns of a conductor wound round the pole core. Current passing through this conductor creates an electromagnet, (rather than the permanent magnets shown in *Figs 1 and 2*).

  (b) a rotating part called the **armature** mounted in bearings housed in the stator and having:

     (iv) a laminated cylinder of iron or steel called the **core**, on which slots are cut to house the

     (v) **armature winding**, i.e. a single or multi-loop conductor system and

    (vi) the **commutator**, (see para 4).

6 The average e.m.f. induced in a single conductor on the armature of a d.c. machine is given by:

$$\frac{\text{flux cut/rev}}{\text{time of 1 rev}} = \frac{2p\Phi}{1/n},$$

where p is the number of **pairs** of poles, Φ is the flux in Wb entering or leaving a pole and n is the speed of rotation in rev/s. Thus the average e.m.f. per conductor is $2p\Phi n$ volts. If there are Z conductors connected in series, the average e.m.f. generated is $2p\Phi nZ$ volts. For a given machine, the number of pairs of poles p and the number of conductors connected in series Z are constant, hence the generated e.m.f. is proportional to Φn. But $2\pi n$ is the angular velocity, ω, in rad/s, hence the generated e.m.f. E is proportional to Φ and to ω, i.e.,

Generated e.m.f., $E \propto \Phi\omega$ (1)

7   The power on the shaft of a d.c. machine is the product of the torque and the angular velocity, i.e.,

Shaft power = $T\omega$ watts,

where T is the torque in Nm and ω is the angular velocity in rad/s. The power developed by the armature current is EI_a watts, where E is the generated e.m.f. in volts and I_a is the armature current in amperes. If losses are neglected then $T\omega = EI_a$. But from para 6, $E \propto \Phi\omega$,

hence $T\omega = \Phi\omega I_a$, i.e., $T \propto \Phi I_a$ $\hspace{2cm}$ (2)

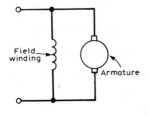
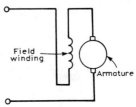

(a) Shunt-wound machine $\hspace{3cm}$ (b) Series-wound machine

Fig 4

8   When the field winding of a d.c. machine is connected in parallel with the armature, as shown in *Fig 4(a)*, the machine is said to be **shunt** wound. If the field winding is connected in series with the armature, as shown in *Fig 4(b)*, then the machine is said to be **series** wound.

9   Depending on whether the electrical machine is series wound or shunt wound, it behaves differently when a load is applied. The behaviour of a d.c. machine under various conditions is shown by means of graphs, called characteristic curves or just **characteristics**. The characteristics shown in the paras below and in the worked problems are theoretical, as they neglect the effects of such things as armature reaction and demagnetising ampere-turns, which are beyond the scope of this text.

10 **Shunt-wound motor characteristics.** The two principal characteristics are the torque/armature current and speed/armature current relationships. From these, the torque/speed relationship can be derived.

(i) The theoretical torque/armature current characteristic can be derived from the expression $T \propto \Phi I_a$, (see para 7). For a shunt-wound motor, the field winding is connected in parallel with the armature circuit and thus the applied voltage gives a constant field current, i.e., a shunt-wound motor is a constant flux machine. Since Φ is constant, it follows that $T \propto I_a$, and the characteristic is as shown in *Fig 5(a)*.

(ii) The armature circuit of a d.c. motor has resistance due to the armature winding and brushes, R_a ohms, and when armature current I_a is flowing through it, there is a voltage drop of $I_a R_a$ volts. In *Fig 5(b)* the armature resistance is shown as a separate resistor in the armature circuit to help understanding. Also, even though the machine is a motor, because conductors are rotating in a magnetic field, a voltage, $E \propto \Phi\omega$, is generated by the armature conductors. By applying Kirchhoff's voltage law to the armature circuit ABCD in *Fig 5(b)*, the voltage equation is $V = E + I_a R_a$, i.e., $E = V - I_a R_a$.

But from para 6, $E \propto \Phi n$, hence $n \propto E/\Phi$, i.e.,

Speed of rotation, $n \propto \dfrac{E}{\Phi} \propto \dfrac{V - I_a R_a}{\Phi}$ $\hspace{2cm}$ (3)

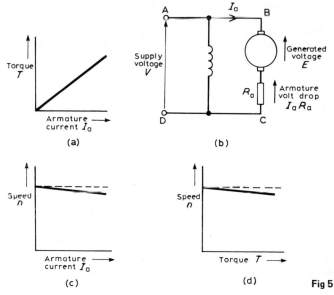

(a)

(b)

(c)

(d)

Fig 5

For a shunt motor, V, Φ and R_a are constants; hence as armature current I_a increases, $I_a R_a$ increases and $V - I_a R_a$ decreases, and the speed is proportional to a quantity which is decreasing and is as shown in *Fig 5(c)*. As the load on the shaft of the motor increases, I_a increases and the speed drops slightly. In practice, the speed falls by about 10% between no-load and full-load on many d.c. shunt-wound motors. Due to this relatively small drop in speed, the d.c. shunt-wound motor is taken as basically being a constant-speed machine and may be used for driving lathes, lines of shafts, drilling machines and so on.

(iii) Since torque is proportional to armature current, (see (i) above), the theoretical speed/torque characteristic is as shown in *Fig 5(d)*.

11 **Series-wound motor characteristics.** The torque/current, speed/current and speed/torque characteristics are discussed in *Problem 7*, page 8 and the characteristics are shown in *Fig 6*.

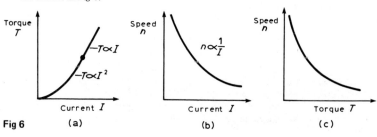

Fig 6

(a)

(b)

(c)

5

B. WORKED PROBLEMS ON D.C. MACHINES

Problem 1 A 200 V d.c. motors develops a shaft torque of 15 Nm at 1200 rev/ min. If the efficiency is 80%, determine the current supplied to the motor.

From para 3, the efficiency of a motor = $\dfrac{\text{output power}}{\text{input power}} \times 100$ per cent

The output power of a motor is the power available to do work at its shaft and is given by $T\omega$ or $(T)(2\pi n)$ watts, where T is the torque in Nm and n is the speed of rotation in rev/s. The input power is the electrical power in watts supplied to the motor, i.e. VI watts. Thus for a motor,

efficiency, $\eta = \left\{ \dfrac{(T)(2\pi n)}{VI} \right\} 100$ per cent

i.e. $80 = \left\{ \dfrac{(15)(2\pi)\left(\dfrac{1200}{60}\right)}{(200)(I)} \right\} (100)$

Thus the current supplied, $I = \dfrac{(15)(2\pi)(20)(100)}{(200)(80)} = \mathbf{11.8\ A}$

Problem 2 A 100 V d.c. generator supplies a current of 15 A when running at 1500 rev/min. If the torque on the shaft driving the generator is 12 Nm, determine (a) the efficiency of the generator and (b) the power loss in the generator.

(a) From para 3, the percentage efficiency of a generator = $\dfrac{\text{output power}}{\text{input power}} \times 100$

The output power is the electrical output, i.e. VI watts. The input power to a generator is the mechanical power in the shaft driving the generator, i.e. $T\omega$ or $(T)(2\pi n)$ watts, where T is the torque in Nm and n is speed of rotation in rev/s. Hence, for a generator,

efficiency, $\eta = \left\{ \dfrac{VI}{(T)(2\pi n)} \right\} 100$

$= \dfrac{(100)(15)(100)}{(12)(2\pi)(\dfrac{1500}{60})}$

$= \mathbf{79.6\ \%}$.

(b) Input power = output power + losses

Hence, $(T)(2\pi n) = VI + $ losses

Losses $= (T)(2\pi n) - VI$

$= (12)(2\pi)(\dfrac{1500}{60}) - (100)(15)$

Power loss $= 1885 - 1500 = \mathbf{385\ watts}$.

Problem 3 A d.c. shunt-wound generator running at constant speed generates a voltage of 150 V at a certain value of field current. Determine the change in the generated voltage when the field current is reduced by 20%, assuming the flux is proportional to the field current.

The generated e.m.f. E of a generator is proportional to $\Phi\omega$, (see para 6), i.e. it is proportional to Φn, where Φ is the flux and n is the speed of rotation. It follows that $E = k\Phi n$, where k is a constant. At speed n_1 and flux Φ_1, $E_1 = k\Phi_1 n_1$. At speed n_2 and flux Φ_2, $E_2 = k\,\Phi_2\,n_2$. Thus, by division:

$$\frac{E_1}{E_2} = \frac{k\,\Phi_1\,n_1}{k\,\Phi_2\,n_2} = \frac{\Phi_1\,n_1}{\Phi_2\,n_2}$$

The initial conditions are $E_1 = 150$ V, $\Phi = \Phi_1$ and $n = n_1$. When the flux is reduced by 20%, the new value of flux is 80/100 or 0.8 of the initial value, i.e., $\Phi_2 = 0.8\Phi_1$. As the generator is running at constant speed, $n_2 = n_1$. Thus:

$$\frac{E_1}{E_2} = \frac{\Phi_1\,n_1}{\Phi_2\,n_2} = \frac{\Phi_1\,n_1}{0.8\,\Phi_1\,n_1} = \frac{1}{0.8}$$

that is, $E_2 = 150 \times 0.8 = 120$ V.

Thus, a reduction of 20% in the value of the flux **reduces the generated voltage to 120 V at constant speed.**

Problem 4 A 200 V d.c. shunt-wound motor has an armature resistance of 0.4 Ω and at a certain load has an armature current of 30 A and runs at 1350 rev/min. If the load on the shaft of the motor is increased so that the armature current increases to 45 A, determine the speed of the motor, assuming the flux remains constant.

The relationship $E \propto \Phi n$ applies to both generators and motors.
For a motor, $E = V - I_a R_a$, (see para 10). Hence

$$E_1 = 200 - (30 \times 0.4) = 188 \text{ V, and}$$
$$E_2 = 200 - (45 \times 0.4) = 182 \text{ V.}$$

With reference to *Problem 3*, the relationship $\dfrac{E_1}{E_2} = \dfrac{\Phi_1\,n_1}{\Phi_2\,n_2}$ applies to both generators and motors. Since the flux is constant, $\Phi_1 = \Phi_2$.

Hence $\dfrac{188}{182} = \dfrac{\Phi_1 \times \dfrac{1350}{60}}{\Phi_1 \times n_2}$, i.e. $n_2 = \dfrac{22.5 \times 182}{188} = 21.78$ rev/s

Thus the speed of the motor when the armature current is 45 A is 21.78×60 rev/min, i.e. **1307 rev/min.**

Problem 5 The shaft torque of a diesel motor driving a 100 V d.c. shunt-wound generator is 25 Nm. The armature current of the generator is 16 A at this value of torque. If the shunt field regulator is adjusted so that the flux is reduced by 15%, the torque increases to 35 Nm. Determine the armature current at this new value of torque.

The shaft torque T of a generator is proportional to ΦI_a, where Φ is the flux and I_a is the armature current. Thus, $T = k\,\Phi\,I_a$ where k is a constant. The torque at flux Φ_1 and armature current I_{a1} is $T_1 = k\,\Phi_1\,I_{a1}$. Similarly,

$$T_2 = k\,\Phi_2\,I_{a2}$$

By division, $\dfrac{T_1}{T_2} = \dfrac{k\,\Phi_1\,I_{a1}}{k\,\Phi_2\,I_{a2}} = \dfrac{\Phi_1\,I_{a1}}{\Phi_2\,I_{a2}}$

Hence $\dfrac{25}{35} = \dfrac{\Phi_1 \times 16}{0.85\Phi_1 \times I_{a2}}$

i.e. $I_{a2} = \dfrac{16 \times 35}{0.85 \times 25} = 26.35$ A

That is, **the armature current at the new value of torque is 26.35 A.**

Problem 6 A 220 V d.c. shunt-wound motor runs at 800 rev/min and the armature current is 30 A. The armature circuit resistance is 0.4 Ω. Determine (a) the maximum value of armature current if the flux is suddenly reduced by 10% and (b) the steady-state value of the armature current at the new value of flux, assuming the shaft torque of the motor remains constant.

(a) For a d.c. shunt-wound motor, $E = V - I_a R_a$. Hence initial generated e.m.f., $E_1 = 220 - (30 \times 0.4) = 208$ V.

The generated e.m.f. is also such that $E \propto \Phi n$, so at the instant the flux is reduced, the speed has not had time to change, and

$$E = 208 \times \frac{90}{100} = 187.2 \text{ V}$$

Hence, the voltage drop due to the armature resistance is $220 - 187.2$, or 32.8 V. The **instantaneous value of the current** is 32.8/0.4, or **82 A**. This increase in current is about three times the initial value and causes an increase in torque, ($T \propto \Phi I_a$). The motor accelerates because of the larger torque value until steady state conditions are reached.

(b) $T \propto \Phi I_a$ and since the torque is constant,
$$\Phi_1 I_{a1} = \Phi_2 I_{a2}. \text{ The flux } \Phi \text{ is reduced by 10\%, hence}$$
$$\Phi_2 = 0.9 \Phi_1.$$
Thus, $\Phi_1 \times 30 = 0.9 \Phi_1 \times I_{a2}$

i.e. the steady state value of armature current, $I_{a2} = \dfrac{30}{0.9} = 33\dfrac{1}{3}$ A.

Problem 7 Sketch the torque/current, speed/current and speed/torque characteristics for a d.c. series-wound motor and with reference to e.m.f. and torque relationships for a d.c. machine, explain their shape.

In a series motor, the armature current flows in the field winding and is equal to the supply current I, (see *Fig 4(b)*).

(i) **The torque/current characteristic**. It is shown in para 7 that torque $T \propto \Phi I_a$. Since the armature and field currents are the same current, I, in a series machine, then $T \propto \Phi I$ over a limited range, before magnetic saturation of the magnetic circuit of the motor is reached, (i.e., the linear portion of the $B-H$ curve for the yoke, poles, air gap, brushes and armature in series). Thus $\Phi \propto I$ and $T \propto I^2$. After magnetic saturation, Φ almost becomes a constant and $T \propto I$. Thus the theoretical torque/current characteristic is as shown in *Fig 6(a)*. A d.c. motor takes a large current on starting and the characteristic shown in *Fig 6(a)* shows that the series-wound motor has a large torque when the current is large. Hence these motors are used for traction (such as trains, milk delivery vehicles etc.), driving fans and for cranes and hoists, where a large initial torque is required.

(ii) **The speed/current characteristic**. It is shown in para 10(ii) that $n \propto (V - I_a R_a)/\Phi$. In a series motor, $I_a = I$ and below the magnetic saturation level, $\Phi \propto I$. Thus $n \propto (V - IR)/I$, where R is the combined resistance of the series field and armature circuit. Since IR is small compared with V, then an approximate relationship for the speed is $n \propto 1/I$. Hence the theoretical speed/current characteristic is as shown in *Fig 6(b)*. The high speed at small values of current indicate that this type of motor must not be run on very light loads and invariably, such motors are permanently coupled to their loads.

(iii) The theoretical speed/torque characteristic may be derived from (i) and (ii) above by obtaining the torque and speed for various values of current and plotting the co-ordinates on the speed/torque characteristic. A typical speed/torque characteristic is shown in *Fig 6(c)*.

C. FURTHER PROBLEMS ON D.C. MACHINES

(a) SHORT ANSWER PROBLEMS

1 An electric . . . . converts electrical energy to mechanical energy.
2 An electric motor converts . . . . energy to . . . . energy.
3 The efficiency of an electrical machine is given by the ratio

$$\frac{\text{. . . . .}}{\text{. . . . .}} \%$$

In *Problems 4 to 7*, an electrical machine runs at n rev/s, has a shaft torque of T and takes a current of I from a supply of voltage V.

4 The power input to a generator is . . . . watts.
5 The power input to a motor is . . . . watts.
6 The power output from a generator is . . . . watts.
7 The power output from a motor is . . . . watts.
8 The generated e.m.f. of a d.c. machine is proportional to . . . . volts.
9 the torque produced by a d.c. motor is proportional to . . . . Nm.
10 In a series-wound d.c. machine, the field winding is in . . . . with the armature circuit.
11 A d.c. motor has its field winding in parallel with the armature circuit. It is called a . . . . wound motor.
12 In a d.c. generator, the relationship between the generated voltage, terminal voltage, current and armature resistance is given by $E = $ . . . .
13 The equation relating the generated e.m.f., terminal voltage, armature current and armature resistance for a d.c. motor is . . . . = ...
14 Since a d.c. shunt motor is essentially a constant . . . . machine, it may be used to drive . . . .
15 A d.c. series motor has a . . . . initial torque and may be used to drive . . . .

(b) MULTI-CHOICE PROBLEMS (answers on page 125)

1 Which of the following statements is false?
   (a) A d.c. motor converts electrical energy to mechanical energy.
   (b) The efficiency of a d.c. motor is the ratio $\dfrac{\text{input power}}{\text{output power}} \times 100$ per cent.
   (c) A d.c. generator converts mechanical energy to electrical energy.
   (d) The efficiency of a d.c. generator is the ratio $\dfrac{\text{output power}}{\text{input power}} \times 100$ per cent.

A shunt-wound d.c. machine is running at n rev/s and has a shaft torque of T Nm. The supply current is I A when connected to d.c. bus bars of voltage V volts. The armature resistance of the machine is R_a ohms, the armature current is I_a A and the generated voltage is E volts. Use this data to find the equations of the

9

quantities stated in *Problems 2 to 9*, selecting the correct answer from the list given below.

(a) $V - I_a R_a$ (b) $E + I_a R_a$ (c) VI

(d) $E - I_a R_a$ (e) $(T)(2\pi n)$ (f) $V + I_a R_a$

2   The input power when running as a generator.

3   The output power when running as a motor.

4   The input power when running as a motor.

5   The ouptut power when running as a generator.

6   The generated voltage when running as a motor.

7   The terminal voltage when running as a generator.

8   The generated voltage when running as a generator.

9   The terminal voltage when running as a motor.

10 Which of the following statements is false?

    (a) A commutator is necessary as part of a d.c. motor to keep the armature rotating in the same direction.

    (b) A commutator is necessary as part of a d.c. generator to produce a uni-directional voltage at the terminals of the generator.

    (c) The field winding of a d.c. machine is housed in slots on the armature.

    (d) The brushes of a d.c. machine are usually made of carbon and do not rotate with the armature.

11 If the speed of a d.c. machine is doubled and the flux remains constant, the generated e.m.f.

    (a) remains the same,   (b) is doubled,   (c) is halved.

12 If the flux per pole of a shunt-wound d.c. generator is increased, and all other variables are kept the same, the speed

    (a) decreases,   (b) stays the same,   (c) increases.

13 If the flux per pole of a shunt-wound d.c. generator is halved, the generated e.m.f. at constant speed

    (a) is doubled   (b) is halved,   (c) remains the same.

14 Which of the following statements is false?

    (a) A series-wound motor has a large starting torque.

    (b) A shunt-wound motor must be permanently connected to its load.

    (c) The speed of a series-wound motor drops considerably when load is applied.

    (d) A shunt-wound motor is essentially a constant speed machine.

(c) CONVENTIONAL PROBLEMS

1   A 250 V, series-wound motor is running at 500 rev/min and its shaft power is 130 Nm. If its efficiency at this load is 88%, find the current taken from the supply.

[30.94 A]

2   In a test on a d.c. motor, the following data was obtained.
Supply voltage: 500 V.   Current taken from the supply: 42.4 A.
Speed: 850 rev/min.   Shaft torque: 187 Nm.
Determine the efficiency of the motor correct to the nearest 0.5%.

[78.5%]

3   The shaft torque required to drive a d.c. generator is 18.7 Nm when it is running at 1250 rev/min. If its efficiency is 87% under these conditions and the armature current is 17.3 A, determine the voltage at the terminals of the generator.

[123.1 V]

4   A 220 V, d.c. generator supplies a load of 37.5 A and runs at 1550 rev/min.

Determine the shaft torque of the diesel motor driving the generator, if the generator efficiency is 78%.

[65.2 Nm]

5 Describe the need for a commutator on the armature of a d.c. generator, using diagrams to illustrate your answer.

6 Explain the function of a commutator on the armature of a d.c. motor.

7 Draw a labelled diagram showing a cross-section of a two-pole d.c. machine. Describe the functions performed by the field windings, the armature, the commutator and the brushes.

8 Determine the generated e.m.f. of a d.c. machine if the armature resistance is 0.1 Ω and it (a) is running as a motor connected to a 230 V supply, the armature current being 60 A, and (b) is running as a generator with a terminal voltage of 230 V, the armature current being 80 A.

[(a) 224 V, (b) 238 V]

9 A d.c. motor has a speed of 900 rev/min when connected to a 460 V supply. Find the approximate value of the speed of the motor when connected to a 200 V supply, assuming the flux decreases by 30% and neglecting the armature voltage drop.

[559 rev/min]

10 A d.c. generator has a generated e.m.f. of 210 V when running at 700 rev/min and the flux per pole is 120 mWb. Determine the generated e.m.f.
(a) at 1050 rev/min, assuming the flux remains constant,
(b) if the flux is reduced by $\frac{1}{6}$ th at constant speed, and
(c) at a speed of 1155 rev/min and a flux of 132 mWb.

[(a) 315 V; (b) 175 V; (c) 381.2 V]

11 A 250 V d.c. shunt-wound generator has an armature resistance of 0.1 Ω. Determine the generated e.m.f. when the generator is supplying 50 kW, neglecting the field current of the generator.

[270 V]

12 A series-wound motor is connected to a d.c. supply and develops full-load torque when the current is 30 A and speed is 1000 rev/min. If the flux per pole is proportional to the current flowing, find the current and speed at half, full-load torque, when connected to the same supply.

[21.2 A; 1415 rev/min]

13 Sketch the theoretical speed/torque characteristic for a series-wound d.c. motor. Use the e.m.f. and torque relationships to explain its shape.

14 Sketch the torque/load characteristic of a series-wound motor and hence give two uses for this type of motor.

15 Using the speed/load characteristic of a shunt-wound machine as the basis of your discussion, give two uses of such a machine.

2 Alternating voltages and currents

A. FORMULAE AND DEFINITIONS ASSOCIATED WITH ALTERNATING VOLTAGES AND CURRENTS

1 Electricity is produced by generators at power stations and then distributed by a vast network of transmission lines (called the National Grid system) to industry and for domestic use. It is easier and cheaper to generate **alternating current** (ac) than direct current (dc) and ac is more conveniently distributed than dc since its voltage can be readily altered using transformers. Whenever dc is needed in preference to ac, devices called rectifiers are used for conversion.

2 Let a single turn coil be free to rotate at constant angular velocity ω symmetrically between the poles of a magnet system as shown in *Fig 1*. An emf is generated in

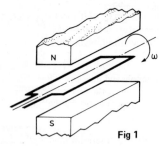

Fig 1

coil (from Faraday's Law) which varies in magnitude and reversed its direction at regular intervals. The reason for this is shown in *Fig 2*.
In positions (a), (e) and (i) the conductors of the loop are effectively moving along the magnetic field, no flux is cut and hence no emf is induced. In position (c) maximum flux is cut and hence maximum emf is induced. In position (g), maximum flux is cut and hence maximum emf is again induced.

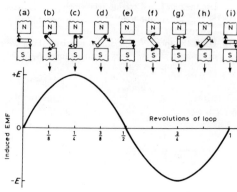

Fig 2

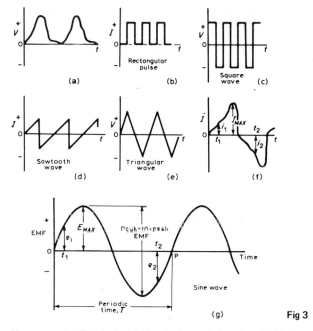

Fig 3

However, using Fleming's right-hand rule, the induced emf is in the opposite direction to that in position (c) and is thus shown as $-E$. In positions (b), (d), (f) and (h) some flux is cut and hence some emf is induced. If all such positions of the coil are considered, in one revolution of the coil, one cycle of alternating emf is produced as shown. This is the principle of operation of the **ac generator** (i.e. the **alternator**).

3 If values of quantities which vary with time t are plotted to a base of time, the resulting graph is called a **waveform**. Some typical waveforms are shown in *Fig 3*. Waveforms (a) and (b) are **unidirectional waveforms**, for, although they vary considerably with time, they flow in one direction only (i.e. they do not cross the time axis and become negative). Waveforms (c) to (g) are called **alternating waveforms** since their quantities are continually changing in direction (i.e. alternately positive and negative).

4 A waveform of the type shown in *Fig 3(g)* is called a **sine wave**. It is the shape of the waveform of emf produced by an alternator and thus the mains electricity supply is of 'sinusoidal' form.

5 One complete series of values is called a **cycle** (i.e. from 0 to P in *Fig 3(g)*).

6 The time taken for an alternating quantity to complete one cycle is called the **period** or the **periodic time**, T, of the waveform.

7 The number of cycles completed in one second is called the **frequency**, f, of the supply and is measured in **hertz**, Hz. The standard frequency of the electricity supply in Great Britain is 50 Hz.

$$T = \frac{1}{f} \quad \text{or} \quad f = \frac{1}{T} .$$

8 **Instantaneous values** are the values of the alternating quantities at any instant of time. They are represented by small letters, i, v, e etc., (see *Figs 3(f)* and *(g)*).

9 The largest value reached in a half cycle is called the **peak value** or the **maximum value** or the **crest value** or the **amplitude** of the waveform. Such values are represented by V_{MAX}, I_{MAX} etc. (see *Figs 3(f)* and *(g)*). A **peak-to-peak** value of emf is shown in *Fig 3(g)* and is the difference between the maximum and minimum values in a cycle.

10 The **average or mean value** of a symmetrical alternating quantity, (such as a sine wave), is the average value measured over a half cycle, (since over a complete cycle the average value is zero).

$$\text{Average or mean value} = \frac{\text{area under the curve}}{\text{length of base}}$$

The area under the curve is found by approximate methods such as the trapezoidal rule, the mid-ordinate rule or Simpson's rule. Average values are represented by V_{AV}, I_{AV}, etc.

For a sine wave, average value = $0.637 \times$ maximum value (i.e. $2/\pi \times$ maximum value).

11 The **effective value** of an alternating current is that current which will produce the same heating effect as an equivalent direct current. The effective value is called the **root mean square (rms) value** and whenever an alternating quantity is given, it is assumed to be the rms value. For example, the domestic mains supply in Great Britain is 240 V and is assumed to mean '240 V rms'. The symbols used for rms values are I, V, E, etc. For a non-sinusoidal waveform as shown in *Fig 4* the rms value is given by:

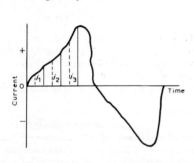

$$I = \sqrt{\left[\frac{i_1^2 + i_2^2 + \ldots + i_n^2}{n}\right]}$$

where n is the number of intervals used.

For a sine wave, rms value = $0.707 \times$ maximum value (i.e. $1/\sqrt{2} \times$ maximum value).

Fig 4

12 (a) Form factor $= \dfrac{\text{rms value}}{\text{average value}}$. For a sine wave, form factor $= 1.11$.

(b) Peak factor $= \dfrac{\text{maximum value}}{\text{rms value}}$. For a sine wave, peak factor $= 1.41$.

The values of form and peak factors give an indication of the shape of waveforms.

13 In *Fig 5*, OA represents a vector that is free to rotate anticlockwise about 0 at an angular velocity of ω rad/s. A rotating vector is known as a **phasor**. After time t seconds the vector OA has turned through an angle ωt. If the line BC is constructed perpendicular to OA as shown, then

$\sin \omega t = \dfrac{BC}{OB}$ i.e. BC = OB $\sin \omega t$.

14

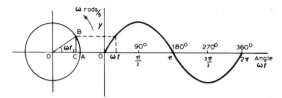

Fig 5

If all such vertical components are projected on to a graph of y against angle ωt (in radians), a sine curve results of maximum value OA. Any quantity which varies sinusoidally can thus be represented as a phasor.

14 A sine curve may not always start at $0°$. To show this a periodic function is represented by $y = \sin(\omega t \pm \phi)$, where ϕ is a phase (or angle) difference compared

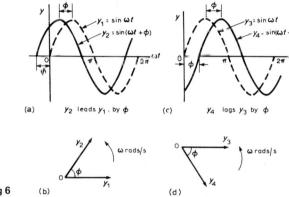

(a) y_2 leads y_1, by ϕ (c) y_4 lags y_3 by ϕ

Fig 6   (b)   (d)

with $y = \sin \omega t$. In *Fig 6(a)*, $y_2 = \sin(\omega t + \phi)$ starts ϕ radians earlier than $y_1 = \sin \omega t$ and is thus said to **lead** y_1 by ϕ radians. Phasors y_1 and y_2 are shown in *Fig 6(b)* at the time when $t = 0$. In *Fig 6(c)*, $y_4 = \sin(\omega t - \phi)$ starts ϕ radians later than $y_3 = \sin \omega t$ and is thus said to **lag** y_3 by ϕ radians. Phasors y_3 and y_4 are shown in *Fig 6(d)* at the time when $t = 0$.

15 Given the general sinusoidal voltage, $v = V_{MAX} \sin(\omega t \pm \phi)$, then

   (i) Amplitude of maximum value $= V_{MAX}$

  (ii) Peak to peak value $= 2V_{MAX}$

 (iii) Angular velocity $= \omega$ rads/s

 (iv) Periodic time, $T = 2\pi/\omega$ seconds.

  (v) Frequency, $f = \omega/2\pi$ Hz (hence $\omega = 2\pi f$).

 (vi) $\phi =$ angle of lag or lead (compared with $v = V_{MAX} \sin \omega t$).

B. WORKED PROBLEMS ON ALTERNATING VOLTAGES AND CURRENTS

(a) FREQUENCY AND PERIODIC TIME

Problem 1 Determine the periodic time for frequencies of (a) 50 Hz and (b) 20 kHz.

(a) Periodic time $T = \dfrac{1}{f} = \dfrac{1}{50}$ = 0.02 s or 20 ms

(b) Periodic time $T = \dfrac{1}{f} = \dfrac{1}{20\,000}$ = 0.000 05 s or 50 μs

Problem 2 Determine the frequencies for periodic times of (a) 4 ms, (b) 4 μs.

(a) Frequency $f = \dfrac{1}{T} = \dfrac{1}{4 \times 10^{-3}} = \dfrac{1000}{4}$ = 250 Hz

(b) Frequency $f = \dfrac{1}{T} = \dfrac{1}{4 \times 10^{-6}} = \dfrac{1\,000\,000}{4}$ = 250 000 Hz or 250 kHz
or 0.25 MHz

Problem 3 An alternating current completes 5 cycles in 8 ms. What is its frequency?

Time for 1 cycle $= \dfrac{8}{5}$ ms = 1.6 ms = periodic time T.

Frequency $f = \dfrac{1}{T} = \dfrac{1}{1.6 \times 10^{-3}} = \dfrac{1000}{1.6} = \dfrac{10\,000}{16}$ = 625 Hz

Further problems on frequency and periodic time may be found in section C(c), Problems 1 to 3, page 23.

(b) AC VALUES OF NON-SINUSOIDAL WAVEFORMS

Problem 4 For the periodic waveforms shown in *Fig 7* determine for each:
(i) frequency; (ii) average value over half a cycle; (iii) rms value; (iv) form factor; and (v) peak factor.

(a) Triangular waveform (Fig 7(a))

(i) Time for 1 complete cycle = 20 ms = periodic time, T.

Hence frequency $f = \dfrac{1}{T} = \dfrac{1}{20 \times 10^{-3}} = \dfrac{1000}{20}$ = 50 Hz

(ii) Area under the triangular waveform for a half cycle

$= \dfrac{1}{2} \times$ base $\times$ height

$= \dfrac{1}{2} \times (10 \times 10^{-3}) \times 200$

$= 1$ volt second.

Average value of waveform

$= \dfrac{\text{area under curve}}{\text{length of base}}$

16

$$= \frac{1 \text{ volt second}}{10 \times 10^{-3} \text{ second}}$$

$$= \frac{1000}{10}$$

$$= 100 \text{ V}$$

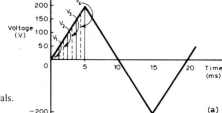

(iii) In *Fig 7(a)*, the first 1/4 cycle is divided into 4 intervals.

Thus rms value

$$= \sqrt{\left[\frac{i_1{}^2 + i_2{}^2 + i_3{}^2 + i_4{}^2}{4}\right]}$$

$$= \sqrt{\left[\frac{25^2 + 75^2 + 125^2 + 175^2}{4}\right]}$$

$$= 114.6 \text{ V}$$

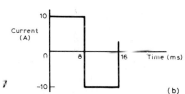

Fig 7

(Note that the greater the number of intervals chosen, the greater the accuracy of the result. For example, if twice the number of ordinates as that chosen above are used, the rms value is found to be 115.6 V)

(iv) Form factor $= \dfrac{\text{rms value}}{\text{average value}} = \dfrac{114.6}{100} = 1.15$

(v) Peak factor $= \dfrac{\text{maximum value}}{\text{rms value}} = \dfrac{200}{114.6} = 1.75$

(b) Rectangular waveform (Fig 7(b))

(i) Time for 1 complete cycle = 16 ms = periodic time, T.

Hence frequency, $f = \dfrac{1}{T} = \dfrac{1}{16 \times 10^{-3}} = \dfrac{1000}{16} = 62.5 \text{ Hz}$

(ii) Average value over half a cycle $= \dfrac{\text{area under curve}}{\text{length of base}}$

$$= \frac{10 \times (8 \times 10^{-3})}{8 \times 10^{-3}} = 10 \text{ A}$$

(iii) The rms value $= \sqrt{\left[\frac{i_1{}^2 + i_2{}^2 + \ldots + i_n{}^2}{n}\right]} = 10 \text{ A}$,

however many intervals are chosen, since the waveform is rectangular.

(iv) Form factor $= \dfrac{\text{rms value}}{\text{average value}} = \dfrac{10}{10} = 1$

(v) Peak factor $= \dfrac{\text{maximum value}}{\text{rms value}} = \dfrac{10}{10} = 1$

Problem 5 The following table gives the corresponding values of current and time for a half cycle of alternating current.

time t (ms)	0	0.5	1.0	1.5	2.0	2.5	3.0	3.5	4.0	4.5	5.0
current i (A)	0	7	14	23	40	56	68	76	60	5	0

Assuming the negative half cycle is identical in shape to the positive half cycle, plot the waveform and find (a) the frequency of the supply, (b) the instantaneous values of current after 1.25 ms and 3.8 ms, (c) the peak or maximum value, (d) the mean or average value, and (e) the rms value of the waveform.

The half cycle of alternating current is shown plotted in *Fig 8*.

(a) Time for a half cycle = 5 ms. Hence the time for 1 cycle, i.e. the periodic time, T = 10 ms or 0.01 s.

   Frequency, $f = \dfrac{1}{T} = \dfrac{1}{0.01} = \textbf{100 Hz}$.

(b) Instantaneous value of current after 1.25 ms is **19 A**, from *Fig 8*.
   Instantaneous value of current after 3.8 ms is **70 A**, from *Fig 8*.

(c) Peak or maximum value = **76A**

(d) Mean or average value = $\dfrac{\text{area under curve}}{\text{length of base}}$

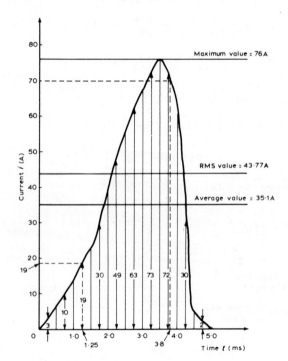

Fig 8

18

Using the mid-ordinate rule with 10 intervals, each of width 0.5 ms gives:

Area under curve $= (0.5 \times 10^{-3})[3+10+19+30+49+63+73+72+30+2]$

(see *Fig 8*)

$$= (0.5 \times 10^{-3})(351)$$

Hence mean or average value $= \dfrac{(0.5 \times 10^{-3})(351)}{5 \times 10^{-3}} = \textbf{35.1 A}$

(e) rms value $= \sqrt{\left[\dfrac{3^2+10^2+19^2+30^2+49^2+63^2+73^2+72^2+30^2+2^2}{10}\right]}$

$$= \sqrt{\left[\dfrac{19\ 157}{10}\right]} = \textbf{43.8 A}$$

Further problems on ac values of non-sinusoidal waveforms may be found in section C(c), Problems 4 to 7, page 24.

(c) AC VALUES OF SINUSOIDAL WAVEFORMS

Problem 6 Calculate the rms value of a sinusoidal current of maximum value 20 A.

For a sine wave,
rms value $= 0.707 \times$ maximum value
$= 0.707 \times 20 = \textbf{14.14 A}$

Problem 7 Determine the peak and mean values for a 240 V mains supply.

For a sine wave, rms value of voltage $V = 0.707 \times V_{MAX}$
A 240 V mains supply means that 240 V is the rms value.

Hence $\qquad V_{MAX} = \dfrac{V}{0.707} = \dfrac{240}{0.707} = \textbf{339.5 V} = \textbf{peak value}$

Mean value $V_{AV} = 0.637 \ V_{MAX} = 0.637 \times 339.5 = \textbf{216.3 V}$

Problem 8 A supply voltage has a mean value of 150 V. Determine its maximum value and its rms value.

For a sine wave, mean value $= 0.637 \times$ maximum value.

Hence maximum value $= \dfrac{\text{mean value}}{0.637} = \dfrac{150}{0.637} = \textbf{235.5 V}$

rms value $= 0.707 \times$ maximum value $= 0.707 \times 235.5 = \textbf{166.5 V}$

Further problems on ac values of sinusoidal waveforms may be found in section C(c), Problems 8 to 12, page 24.

(d) $v = V_{MAX} \sin(\omega t \pm \phi)$

Problem 9 An alternating voltage is given by $v = 282.8 \sin 314\, t$ volts. Find (a) the rms voltage, (b) the frequency and (c) the instantaneous value of voltage when $t = 4$ ms.

(a) The general expression for an alternating voltage is $v = V_{MAX} \sin(\omega t \pm \phi)$.
Comparing $v = 282.8 \sin 314\, t$ with this general expression gives the peak voltage as 282.8 V
Hence the rms voltage = 0.707 × maximum value = 0.707 × 282.8 = **200 V**

(b) Angular velocity, $\omega = 314$ rads/s i.e. $2\pi f = 314$

Hence frequency, $f = \dfrac{314}{2\pi} = $ **50 Hz**

(c) When $t = 4$ ms, $v = 282.8 \sin(314 \times 4 \times 10^{-3})$
$$= 282.8 \sin(1.256)$$

1.256 radians $= \left(1.256 \times \dfrac{180}{\pi}\right)^{\circ} = 71.96^{\circ} = 71^{\circ}\ 58'$

Hence $v = 282.8 \sin 71^{\circ}\ 58'$
$$= \textbf{268.9 V}$$

Problem 10 An alternating voltage is given by $v = 75 \sin(200\pi t - 0.25)$ volts. Find (a) the amplitude; (b) the peak-to-peak value; (c) the rms value, (d) the periodic time; (e) the frequency; and (f) the phase angle (in degrees and minutes) relative to $75 \sin 200\pi t$.

Comparing $v = 75 \sin(200\pi t - 0.25)$ with the general expression
$v = V_{MAX} \sin(\omega t \pm \phi)$ gives:
(a) Amplitude, or peak value = **75 V**
(b) Peak-to-peak value = 2 × 75 = **150 V**
(c) The rms value = 0.707 × maximum value = 0.707 × 75 = **53 V**
(d) Angular velocity, $\omega = 200\pi$ rads/s

Hence periodic time, $T = \dfrac{2\pi}{\omega} = \dfrac{2\pi}{200\pi} = \dfrac{1}{100} = $ **0.01 s or 10 ms**

(e) Frequency, $f = \dfrac{1}{T} = \dfrac{1}{0.01} = $ **100 Hz**

(f) Phase angle, $\phi = 0.25$ radians lagging $75 \sin 200\pi t$

0.25 rads $= \left(0.25 \times \dfrac{180}{\pi}\right)^{\circ} = 14.32^{\circ} = 14^{\circ}\ 19'$

Hence phase angle = **14° 19' lagging**

Problem 11 An alternating voltage, v, has a periodic time of 0.01 s and a peak value of 40 V. When time t is zero, $v = -20$ V. Express the instantaneous voltage in the form $v = V_{MAX} \sin(\omega t + \phi)$ V.

Amplitude, $V_{MAX} = 40$ V.

Periodic time $T = \dfrac{2\pi}{\omega}$. Hence angular velocity, $\omega = \dfrac{2\pi}{T} = \dfrac{2\pi}{0.01} = 200\pi$ rads/s

$v = V_{MAX} \sin{(\omega t + \phi)}$ thus becomes $v = 40 \sin{(200\pi t + \phi)}$ V.

When time $t = 0$, $v = -20$ V

i.e. $-20 = 40 \sin \phi$

$\sin \phi = \dfrac{-20}{40} = -0.5$

Hence $\phi = \arcsin{-0.5} = -30° = \left(-30 \times \dfrac{\pi}{180}\right)$ rads $= -\dfrac{\pi}{6}$ rads

Thus $v = 40 \sin{(200\pi t - \dfrac{\pi}{6})}$ V

Problem 12 The current in an ac circuit at any time t seconds is given by.
$i = 120 \sin{(100\pi t + 0.36)}$ amperes. Find:
(a) the peak value, the periodic time, the frequency and phase angle relative to $120 \sin 100\pi t$;
(b) the value of the current when $t = 0$;
(c) the value of the current when $t = 8$ ms;
(d) the time when the current first reaches 60 A, and
(e) the time when the current is first a maximum.

(a) Peak value = **120 A.**

Periodic time $T = \dfrac{2\pi}{\omega} = \dfrac{2\pi}{100\pi}$ (since $\omega = 100\pi$)

$= \dfrac{1}{50} = $ **0.02 s or 20 ms**

Frequency $f = \dfrac{1}{T} = \dfrac{1}{0.02} = $ **50 Hz**

Phase angle $= 0.36$ rads $= \left(0.36 \times \dfrac{180}{\pi}\right)° = 20° \ 38'$ **leading**

(b) When $t = 0$, $i = 120 \sin{(0+0.36)} = 120 \sin 20° \ 38' = $ **42.29 A**

(c) When $t = 8$ ms, $i = 120 \sin\left[100\pi \left(\dfrac{8}{10^3}\right) +0.36\right]$

$= 120 \sin 2.8733$
$= 120 \sin 164° \ 38'$
$= $ **31.80 A**

(d) When $i = 60$ A, $60 = 120 \sin{(100\pi t +0.36)}$

$\dfrac{60}{120} = \sin{(100\pi t+0.36)}$

$(100\pi t+0.36) = \arcsin 0.5 = 30° = \dfrac{\pi}{6}$ rads $= 0.5236$ rads

Hence time, $t = \dfrac{0.5236-0.36}{100\pi} = $ **0.5208 ms**

(e) When the current is a maximum, $i = 120$ A

Thus $120 = 120 \sin(100\pi t + 0.36)$

$1 = \sin(100\pi t + 0.36)$

$(100\pi t + 0.36) = \arcsin 1 = 90° = \dfrac{\pi}{2}$ rads $= 1.5708$ rads

Hence time, $t = \dfrac{1.5708 - 0.36}{100\pi} = 3.854$ ms

Further problems on the general expression $v = V_{MAX} \sin(\omega t \pm \phi)$ may be found in section C(c), Problems 13 to 18, page 25.

C. FURTHER PROBLEMS ON ALTERNATING VOLTAGES AND CURRENTS

(a) SHORT ANSWER PROBLEMS

1 Briefly explain the principle of the simple alternator.
2 What is the difference between an alternating and a unidirectional waveform.
3 What is meant by (a) waveform; (b) cycle.
4 The time to complete one cycle of a waveform is called the . . . .
5 What is frequency? Name its unit.
6 The mains supply voltage has a special shape of waveform called a . . . .
7 Define peak value.
8 What is meant by the rms value?
9 The domestic mains electricity supply voltage in Great Britain is . . . .
10 What is the mean value of a sinusoidal alternating emf which has a maximum value of 100 V?
11 The effective value of a sinusoidal waveform is . . . . $\times$ maximum value.
12 What is a phasor quantity?
13 Complete the statement: Form factor $= \dfrac{\cdots\cdots\cdots\cdots}{\cdots\cdots\cdots\cdots}$, and for a sine wave,

form factor $= \dfrac{\cdots\cdots\cdots\cdots}{\cdots\cdots\cdots\cdots}$.

14 Complete the statement: Peak factor $= \dfrac{\cdots\cdots\cdots\cdots}{\cdots\cdots\cdots\cdots}$, and for a sine wave,

peak factor $= \dfrac{\cdots\cdots\cdots\cdots}{\cdots\cdots\cdots\cdots}$.

15 A sinusoidal current is given by $i = I_{MAX} \sin(\omega t \pm \alpha)$. What do the symbols I_{MAX}, ω and α represent?

(b) MULTI-CHOICE PROBLEMS (answers on page 125)

1 The value of an alternating current at any given instant is (a) a maximum value; (b) a peak value; (c) an instantaneous value; (d) an rms value.
2 An alternating current completes 100 cycles in 0.2 s. Its frequency is (a) 20 Hz; (b) 100 Hz; (c) 0.002 Hz; (d) 1 kHz.
3 In *Fig 9*, at the instant shown the generated emf will be (a) zero; (b) an rms value; (c) an average value; (d) a maximum value.
4 The supply of electrical energy for a consumer is usually by ac because
(a) transmission and distribution are more easily effected;
(b) it is most suitable for variable speed motors;
(c) the volt drop in cables is minimal; (d) cable power losses are negligible.

22

5　Which of the following statements is false?
　(a) It is cheaper to use ac than dc.
　(b) Distribution of ac is more convenient than with dc since voltages may be readily altered using transformers.
　(c) An alternator is an ac generator.
　(d) A rectifier changes dc into ac.

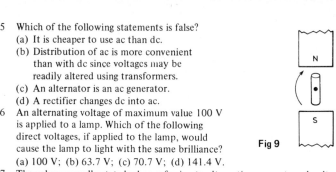

Fig 9

6　An alternating voltage of maximum value 100 V is applied to a lamp. Which of the following direct voltages, if applied to the lamp, would cause the lamp to light with the same brilliance?
　(a) 100 V; (b) 63.7 V; (c) 70.7 V; (d) 141.4 V.

7　The value normally stated when referring to alternating currents and voltages is the
　(a) instantaneous value; (b) rms value; (c) average value; (d) peak value.

8　State which of the following is false. For a sine wave:
　(a) the peak factor is 1.414.
　(b) the rms value is 0.707 × peak value.
　(c) the average value is 0.637 × rms value.
　(d) the form factor is 1.11.

9　An ac supply is 70.7 V, 50 Hz. Which of the following statements is false?
　(a) The periodic time is 20 ms.
　(b) The peak value of the voltage is 70.7 V.
　(c) The rms value of the voltage is 70.7 V.
　(d) The peak value of the voltage is 100 V.

10　An alternating voltage is given by $v = 100 \sin (50\pi t - 0.30)$ V. Which of the following statements is true?
　(a) The rms voltage is 100 V;
　(b) The periodic time is 20 ms;
　(c) The frequency is 25 Hz;
　(d) The voltage is leading $v = 100 \sin 50\pi t$ by 0.30 radians.

(c) CONVENTIONAL PROBLEMS

Frequency and periodic time

1　Determine the periodic time for the following frequencies:
　(a) 2.5 Hz; (b) 100 Hz; (c) 40 kHz.

[(a) 0.4 s; (b) 10 ms; (c) 25 μs]

2　Calculate the frequency for the following periodic times:
　(a) 5 ms; (b) 50 μs; (c) 0.2 s.

[(a) 0.2 kHz; (b) 20 kHz; (c) 5 Hz]

3　An alternating current completes 4 cycles in 5 ms. What is its frequency?

[800 Hz]

a.c. values of non-sinusoidal waveforms

4　An alternating current varies with time over half a cycle as follows;

Current (A)	0	0.7	2.0	4.2	8.4	8.2	2.5	1.0	0.4	0.2	0
time (ms)	0	1	2	3	4	5	6	7	8	9	10

The negative half cycle is similar. Plot the curve and determine:
(a) the frequency; (b) the instantaneous values at 3.4 ms and 5.8 ms; (c) its mean value and (d) its rms value.

[(a) 50 Hz. (b) 5.5 A, 3.4 A; (c) 2.8 A; (d) 4.0 A]

23

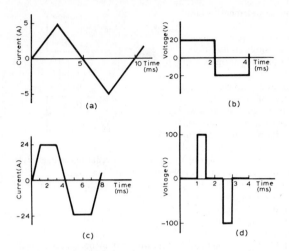

Fig 10

5 For the waveforms shown in *Fig 10* determine for each (i) the frequency; (ii) the average value over half a cycle; (iii) the rms value; (iv) the form factor; (v) the peak factor.

$$\begin{bmatrix} \text{(a)} & \text{(i) 100 Hz;} & \text{(ii) 2.50A;} & \text{(iii) 2.88A;} & \text{(iv) 1.15;} & \text{(v) 1.74} \\ \text{(b)} & \text{(i) 250 Hz;} & \text{(ii) 20 V;} & \text{(iii) 20 V;} & \text{(iv) 1.0;} & \text{(v) 1.0} \\ \text{(c)} & \text{(i) 125 Hz;} & \text{(ii) 18 A;} & \text{(iii) 19.56 A;} & \text{(iv) 1.09;} & \text{(v) 1.23} \\ \text{(d)} & \text{(i) 250 Hz;} & \text{(ii) 25 V;} & \text{(iii) 50 V;} & \text{(iv) 2.0;} & \text{(v) 2.0} \end{bmatrix}$$

6 An alternating voltage is triangular in shape, rising at a constant rate to a maximum of 300 V in 8 ms and then falling to zero at a constant rate in 4 ms. The negative half cycle is identical in shape to the positive half cycle. Calculate (a) the mean voltage over half a cycle, and (b) the rms voltage.

[(a) 150 V; (b) 171 V]

7 An alternating emf varies with time over half a cycle as follows:

emf (V)	0	45	80	155	215	320	210	95	0
time (ms)	0	1.5	3.0	4.5	6.0	7.5	9.0	10.5	12.0

The negative half cycle is identical in shape to the positive half cycle. Plot the waveform and determine (a) the periodic time and frequency; (b) the instantaneous value of voltage at 3.75 ms, (c) the times when the voltage is 125 V; (d) the mean value, and (e) the rms value.

[(a) 24 ms, 41.67 Hz; (b) 115 V; (c) 4 ms and 10.1 ms; (d) 142 V; (e) 171 V]

a.c. values of sinusoidal waveforms

8 Calculate the rms value of a sinusoidal curve of maximum value 300 V.

[212.1 V]

9 Find the peak and mean values for a 200 V mains supply.

[282.9 V; 180.2 V]

10 Plot a sine wave of peak value 10.0 A. Show that the average value of the waveform is 6.37 A over half a cycle, and that the rms value is 7.07 A.

11 A sinusoidal voltage has a maximum value of 120 V. Calculate its rms and average values.

[84.8 V; 76.4 V]

12 A sinusoidal current has a mean value of 15.0 A. Determine its maximum and rms values.

[23.55 A; 16.65 A]

$v = V_{MAX} \sin (\omega t \pm \phi)$

13 An alternating voltage is represented by $v = 20 \sin 157.1\ t$ volts. Find (a) the maximum value; (b) the frequency; and (c) the periodic time. (d) What is the angular velocity of the phasor representing this waveform?

[(a) 20 V; (b) 25 Hz; (c) 0.04 s; (d) 157.1 rads/s]

14 Find the peak value, the rms value, the periodic time, the frequency and the phase angle (in degrees and minutes) of the following alternating quantities:
(a) $v = 90 \sin 400\pi t$ volts.   [(a) 90 V; 63.63 V; 5 ms; 200 Hz; 0°]
(b) $i = 50 \sin (100\pi t + 0.30)$ amperes.

[(b) 50 A; 35.35 A; 0.02 s; 50 Hz; 17°11' lead]

(c) $e = 200 \sin (628.4t - 0.41)$ volts.

[(c) 200 V; 141.4 V; 0.01 s; 100 Hz; 23°29' lag]

15 A sinusoidal current has a peak value of 30 A and a frequency of 60 Hz. At time $t = 0$, the current is zero. Express the instantaneous current i in the form $i = I_{MAX} \sin \omega t$.

[$i = 30 \sin 120\pi t$]

16 An alternating voltage v has a periodic time of 20 ms and a maximum value of 200 V When time $t = 0$, $v = -75$ volts. Deduce a sinusoidal expression for v and sketch one cycle of the voltage showing important points.

[$v = 200 \sin (100\pi t - 0.384)$]

17 The voltage in an alternating current circuit at any time t seconds is given by $v = 60 \sin 40t$ volts. Find the first time when the voltage is (a) 20 V, and (b) −30 V.

[(a) 8.496 ms; (b) 91.63 ms]

18 The instantaneous value of voltage in an ac circuit at any time t seconds is given by $v = 100 \sin (50\pi t - 0.523)$ V. Find:
(a) the peak-to-peak voltage, the periodic time, the frequency and the phase angle;
(b) the voltage when $t = 0$;   (c) the voltage when $t = 8$ ms;
(d) the times in the first cycle when the voltage is 60 V;
(e) the times in the first cycle when the voltage is −40 V; and
(f) the first time when the voltage is a maximum.
Sketch the curve for one cycle showing relevant points.

[(a) 200 V, 0.04 s, 25 Hz, 29°58' lagging; (b) −49.95 V;
(c) 66.96 V; (d) 7.426 ms, 19.23 ms; (e) 25.95 ms, 40.71 ms;
(f) 13.33 ms.]

3 Effects of inductance and capacitance in a.c. circuits

A. MAIN POINTS CONCERNED WITH THE EFFECTS OF INDUCTANCE AND CAPACITANCE IN A.C. CIRCUITS

(a) INDUCTANCE

1 In a purely resistive a.c. circuit, the current I_R and applied voltage V_R are in phase. See *Fig 1(a)*.

2 In a purely inductive a.c. circuit, the current I_L lags the applied voltage V_L by 90° (i.e. $\pi/2$ rads). See *Fig 1(b)*.

3 In a purely inductive circuit the opposition to the flow of alternating current is called the inductive reactance, X_L.

$$X_L = \frac{V_L}{I_L} = 2\pi fL \text{ ohms, where } f \text{ is the supply frequency in hertz and } L \text{ is the inductance in henry's.}$$

X_L is proportional to f, as shown in *Fig 2*.

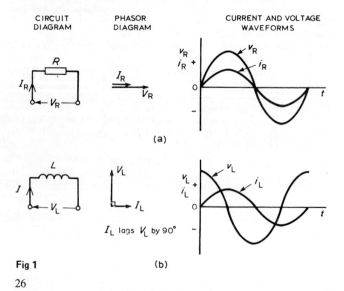

Fig 1

(b)

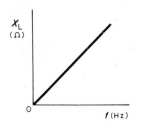

Fig 2

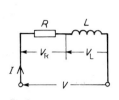

CIRCUIT DIAGRAM

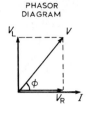

PHASOR DIAGRAM

Fig 3

4   In an a.c. series circuit containing inductance L and resistance R, the applied voltage V is the phasor sum of V_R and V_L (see *Fig 3*), and thus the current I lags the applied voltage V by an angle lying between $0°$ and $90°$ (depending on the values of V_R and V_L), shown as angle ϕ. In any a.c. series circuit, the current is common to each component and is thus taken as a reference phasor.

5   In an a.c. circuit, the ratio $\dfrac{\text{applied voltage } V}{\text{current } I}$ is called the impedance Z,

   i.e. $Z = \dfrac{V}{I}$ ohms.

6   From the phasor diagram of *Fig 3*, the **voltage triangle** is derived, as shown in *Fig 4(a)*.
   For the $R-L$ circuit: $V = \sqrt{(V_R{}^2 + V_L{}^2)}$ (by Pythagoras' theorem)
   and $\tan \phi = \dfrac{V_L}{V_R}$ (by trigonometric ratios)

7   If each side of the voltage triangle in *Fig 4(a)* is divided by current I then the **impedance triangle** is derived, as shown in *Fig 4(b)*.
   For the $R-L$ circuit: $Z = \sqrt{(R^2 + X_L{}^2)}$
   $\tan \phi = \dfrac{X_L}{R}$, $\sin \phi = \dfrac{X_L}{Z}$ and $\cos \phi = \dfrac{R}{Z}$
   (See *Problems 1 to 7*).

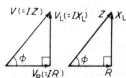

VOLTAGE TRIANGLE   IMPEDANCE TRIANGLE

Fig 4

(a) VOLTAGE TRIANGLE   (b) IMPEDANCE TRIANGLE

(b) CAPACITANCE

8   In a purely capacitive a.c. circuit, the current I_C leads the applied voltage V_C by $90°$ (i.e., $\pi/2$ rads). See *Fig 5*.

9   In a purely capacitive circuit the opposition to the flow of alternating current is called the capacitive reactance, X_C.

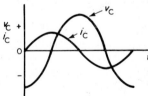

Fig 5 I_C leads V_C by $90°$

27

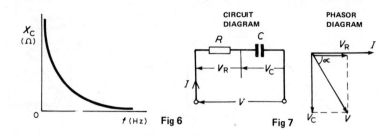

CIRCUIT DIAGRAM   **PHASOR DIAGRAM**

Fig 6   Fig 7

$X_C = \dfrac{V_C}{I_C} = \dfrac{1}{2\pi fC}$ ohms, where C is the capacitance in farads. X_C varies with f as shown in *Fig 6*.

10 In an a.c. series circuit containing capacitance C and resistance R, the applied voltage V is the phasor sum of V_R and V_C (see *Fig 7*) and thus the current I leads the applied voltage V by an angle lying between $0°$ and $90°$, (depending on the values of V_R and V_C), shown as angle α.

11 From the phasor diagram of *Fig 7*, the voltage triangle is derived as shown in *Fig 8(a)*.

For the $R-C$ circuit: $V = \sqrt{(V_R^2 + V_C^2)}$

and $\tan \alpha = \dfrac{V_C}{V_R}$

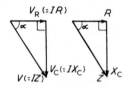

12 If each side of the voltage triangle in *Fig 8(a)* divided by current I then the impedance triangle is derived, as shown in *Fig 8(b)*.

For the $R-C$ circuit: $Z = \sqrt{(R^2 + X_C^2)}$

$\tan \alpha = \dfrac{X_C}{R}$, $\sin \alpha = \dfrac{X_C}{Z}$ and $\cos \alpha = \dfrac{R}{Z}$

(a) **VOLTAGE TRIANGLE**   (b) **IMPEDANCE TRIANGLE**

Fig 8

(See *Problems 8 to 12*).

B. WORKED PROBLEMS ON THE EFFECTS OF INDUCTANCE AND CAPACITANCE IN A.C. CIRCUITS

(a) INDUCTANCE

Problem 1 (a) Calculate the reactance of a coil of inductance 0.32 H when it is connected to a 50 Hz supply.

(b) A coil has a reactance of 124 ohms in a circuit with a supply of frequency 5 kHz. Determine the inductance of the coil.

(a) Inductive reactance, $X_L = 2\pi fL = 2\pi(50)(0.32) = \mathbf{100.5\ \Omega}$

(b) Since $X_L = 2\pi fL$, inductance $L = \dfrac{X_L}{2\pi f} = \dfrac{124}{2\pi(5000)}$ H = **3.95 mH**.

Problem 2 A coil has an inductance of 40 mH and negligible resistance. Calculate its inductive reactance and the resulting current if connected to (a) a 240 V, 50 HZ supply, (b) a 100 V, 1 kHz supply.

(a) Inductive reactance $X_L = 2\pi f L = 2\pi(50)(40 \times 10^{-3}) = $ **12.57 Ω**.

$$\text{Current } I = \frac{V}{X_L} = \frac{240}{12.57} = \textbf{19.09 A}$$

(b) Inductive reactance $X_L = 2\pi(1000)(40 \times 10^{-3}) = $ **251.3 Ω**

$$\text{Current } I = \frac{V}{X_L} = \frac{100}{251.3} = \textbf{0.398 A}$$

Problem 3 In a series $R-L$ circuit, the p.d. across the resistance R is 12 V and the p.d. across the inductance L is 5 V. Find the supply voltage and the phase angle between current and voltage.

From the voltage triangle of *Fig 4(a)*, supply voltage $V = \sqrt{(12^2 + 5^2)}$
$$= \textbf{13 volts}$$
(Note that in a.c. circuits, the supply voltage is **not** the arithmetic sum of the p.d.'s across components. It is, in fact, the **phasor sum**.)

$\tan \phi = \dfrac{V_L}{V_R} = \dfrac{5}{12}$, from which $\phi = \arctan \dfrac{5}{12} - $ **22° 37' lagging**.

('Lagging' infers that the current is 'behind' the voltage, since phasors revolve anticlockwise.)

Problem 4 A coil has a resistance of 12 ohms and an inductance of 15.9 mH. Calculate (a) the reactance, (b) the impedance, and (c) the current taken from a 240 V, 50 Hz supply. Determine also the phase angle between the supply voltage and current.

$R = 12\Omega$; $L = 15.9$ mH $= 15.9 \times 10^{-3}$ H; $f = 50$ Hz; $V = 240$ V.
(a) Inductive reactance $X_L = 2\pi f L = 2\pi(50)(15.9 \times 10^{-3}) = 5\Omega$
(b) Impedance $Z = \sqrt{(R^2 + X_L^2)} = \sqrt{(12^2 + 5^2)}$ $= 13\Omega$
(c) Current $I = \dfrac{V}{Z} = \dfrac{240}{13} = $ **18.5 A**

The circuit and phasor diagrams and the voltage and impedance triangles are as shown in *Figs 3 and 4*.

Since $\tan \phi = \dfrac{X_L}{R}$, $\phi = \arctan \dfrac{X_L}{R} = \arctan \dfrac{5}{12} = $ **22° 37' lagging**.

Problem 5 A coil takes a current of 2 A from a 12 V d.c. supply. When connected to a 240 V, 50 Hz a.c. supply the current is 20 A. Calculate the resistance, impedance, inductive reactance and inductance of the coil.

Resistance $R = \dfrac{\text{d.c. voltage}}{\text{d.c. current}} = \dfrac{12}{2} = 6\Omega$

Impedance $Z = \dfrac{\text{a.c. voltage}}{\text{a.c. current}} = \dfrac{240}{20} = 12\Omega$

As $Z = \sqrt{(R^2 + X_L^2)}$, inductive reactance, $X_L = \sqrt{(Z^2 - R^2)}$
$$= \sqrt{(12^2 - 6^2)} = 10.39\Omega$$
As $X_L = 2\pi f L$, inductance $L = \dfrac{X_L}{2\pi f} = \dfrac{10.39}{2\pi(50)} = $ **33.1 mH**

This problem indicates a simple method for finding the inductance of a coil, i.e. firstly to measure the current when the coil is connected to a d.c. supply of known voltage, and then to repeat the process with an a.c. supply.

Problem 6 A coil of inductance 318.3 mH and negligible resistance is connected in series with a 200Ω resistor to a 240 V, 50 Hz supply. Calculate (a) the inductive reactance of the coil, (b) the impedance of the circuit, (c) the current in the circuit, (d) the p.d. across each component, and (e) the circuit phase angle.

L = 318.3 mH = 0.3183 H; R = 200Ω; V = 240 V; f = 50 Hz.
The circuit diagram is as shown in *Fig 3*.
(a) Inductive reactance $X_L = 2\pi fL = 2\pi(50)(0.3183) =$ **100Ω**
(b) Impedance $Z = \sqrt{(R^2 + X_L^2)} = \sqrt{[(200)^2 + (100)^2]} =$ **223.6Ω**
(c) Current $I = \dfrac{V}{Z} = \dfrac{240}{223.6} =$ **1.073 A.**
(d) The p.d. across the coil, $\quad V_L = I X_L = 1.073 \times 100 =$ **107.3 V**
    The p.d. across the resistor, $V_R = I R \quad = 1.073 \times 200 =$ **214.6 V**
    [Check: $\sqrt{(V_R^2 + V_L^2)} = \sqrt{[(214.6)^2 + (107.3)^2]} = 240$ V, the supply voltage]
(e) From the impedance triangle, angle $\phi = \arctan \dfrac{X_L}{R} = \arctan \dfrac{100}{200}$

    Hence **phase angle Φ = 26° 34′ lagging**

Problem 7 A coil has a resistance of 100 Ω and an inductance of 200 mH. If an alternating voltage v, given by $v = 200 \sin 500\,t$ volts is applied across the coil, calculate (a) the circuit impedance, (b) the current flowing, (c) the p.d. across the resistance, (d) the p.d. across the inductance and (e) the phase angle between voltage and current.

Because $v = 200 \sin 500\,t$ volts, $V_{MAX} = 200$ V and $\omega = 2\pi f = 500$ rad/s.
Hence r.m.s. voltage $V = 0.707 \times 200 = 141.4$ V.
Inductive reactance $X_L = 2\pi fL = \omega L = 500 \times 200 \times 10^{-3} = 100$Ω
(a) Impedance $Z = \sqrt{(R^2 + X_L^2)} = \sqrt{(100^2 + 100^2)} =$ **141.4Ω**
(b) Current $I = \dfrac{V}{Z} = \dfrac{141.4}{141.4} =$ **1A.**
(c) The p.d. across the resistance $V_R = IR = 1 \times 100 =$ **100 V**
(d) P.D. across the inductance, $V_L = I X_L = 1 \times 100 =$ **100 V.**
(e) Phase angle between voltage and current, $\phi = \arctan \dfrac{X_L}{R} = \arctan \dfrac{100}{100}$

Hence $\phi = 45°$ or $\dfrac{\pi}{4}$ rads

Further problems on the effects of inductance in a.c. circuits may be found in section C(c), Problems 1 to 11, page 33.

(b) CAPACITANCE

Problem 8 Determine the capacitive reactance of a capacitor of 10 μF when connected to a circuit of frequency (a) 50 Hz, (b) 20 kHz.

(a) Capacitive reactance $X_C = \dfrac{1}{2\pi fC} = \dfrac{1}{2\pi(50)(10 \times 10^{-6})} = \dfrac{10^6}{2\pi(50)(10)}$
    $= $ **318.3Ω**

(b) $X_C = \dfrac{1}{2\pi fC} = \dfrac{1}{2\pi(20 \times 10^3)(10 \times 10^{-6})} = \dfrac{10^6}{2\pi(20 \times 10^3)(10)} =$ **0.796Ω**

Hence as the frequency is increased from 50 Hz to 20 kHz, X_C decreases from 318.3Ω to 0.796Ω (see *Fig 6*).

Problem 9 A capacitor has a reactance of 40Ω when operated on a 50 Hz supply. Determine the value of its capacitance.

$$X_C = \frac{1}{2\pi fC} \text{, so capacitance } C = \frac{1}{2\pi fX_C} = \frac{1}{2\pi(50)(40)} \text{ F}$$

$$= \frac{10^6}{2\pi(50)(40)} \mu F = 79.58\mu F$$

Problem 10 Calculate the current taken by a 23 μF capacitor when connected to a 240 V, 50 Hz supply.

$$\text{Current } I = \frac{V}{X_C} = \frac{V}{\frac{1}{2\pi fC}} = 2\pi fCV = 2\pi(50)(23 \times 10^{-6})(240)$$

$$= 1.73 \text{ A}$$

Problem 11 A resistor of 25Ω is connected in series with a capacitor of 45μF. Calculate (a) the impedance, and (b) the current taken from a 240 V, 50 Hz supply. Find also the phase angle between the supply voltage and the current.

$R = 25\Omega$; $C = 45\mu F = 45 \times 10^{-6}$ F; $V = 240$ V; $f = 50$ Hz.
The circuit diagram is as shown in *Fig 7*.

Capacitive reactance $X_C = \frac{1}{2\pi fC} = \frac{1}{2\pi(50)(45 \times 10^{-6})} = 70.74\Omega$

(a) Impedance $Z = \sqrt{(R^2 + X_C^2)} = \sqrt{[(25)^2 + (70.74)^2]} = 75.03\Omega$

(b) Current $I = \frac{V}{Z} = \frac{240}{75.03} = 3.20 \text{ A}$

Phase angle between the supply voltage and current, $\alpha = \arctan \frac{X_C}{R}$

Hence $\alpha = \arctan \frac{70.74}{25} = 70° 32'$ **leading**.

('Leading' infers that the current is 'ahead' of the voltage, as phasors revolve anticlockwise.)

Problem 12 A capacitor C is connected in series with a 40Ω resistor across a supply of frequency 60 Hz. A current of 3 A flows and the circuit impedance is 50Ω. Calculate (a) the value of capacitance C, (b) the supply voltage, (c) the phase angle between the supply voltage and current, (d) the p.d. across the resistor, and (e) the p.d. across the capacitor. Draw the phasor diagram.

(a) Impedance $Z = \sqrt{(R^2 + X_C^2)}$

Hence $X_C = \sqrt{(Z^2 - R^2)} = \sqrt{(50^2 - 40^2)} = 30\Omega$

$X_C = \frac{1}{2\pi fC}$. Hence $C = \frac{1}{2\pi fX_C} = \frac{1}{2\pi(60)30}$ F = **88.42 μF**

(b) Since $Z = \frac{V}{I}$ then $V = IZ = (3)(50) = 150$ V

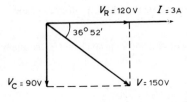

$V_R = 120\,V \qquad I = 3A$

$36°\,52'$

$V_C = 90V \qquad V = 150V$

Phasor diagram   **Fig 9**

  (c) Phase angle $\alpha = \arctan \dfrac{X_C}{R} = \arctan \dfrac{30}{40} = \mathbf{36°\ 52'\ leading}$

  (d) p.d. across resistor $V_R = IR = (3)(40) = \mathbf{120\ V}$

  (e) p.d. across capacitor, $V_C = IX_C = (3)(30) = \mathbf{90\ V}$

    The phasor diagram is shown in *Fig 9*.

Further problems on the effects of capacitance in a.c. circuits may be found in section C(c), Problems 12 to 20, page 34.

C. FURTHER PROBLEMS ON THE EFFECTS OF INDUCTANCE AND CAPACITANCE IN A.C. CIRCUITS

(a) SHORT ANSWER PROBLEMS

1 Complete the following statements:
    (a) In a purely resistive a.c. circuit the current is . . . . . . with the voltage.
    (b) In a purely inductive a.c. circuit the current . . . . the voltage by . . . . .
    degrees.
    (c) In a purely capacitive a.c. circuit the current . . . . . . the voltage by . . . . .
    degrees.

2 Draw phasor diagrams to represent (a) a purely resistive a.c. circuit, (b) a purely inductive a.c. circuit, and (c) a purely capacitive a.c. circuit.

3 What is inductive reactance? State the symbol and formula for determining inductive reactance.

4 What is capacitive reactance? State the symbol and formula for determining capacitive reactance.

5 What does 'impedance' mean when referring to an a.c. circuit?

6 Draw an impedance triangle for an $R-L$ circuit. Derive from the triangle an expression for (a) impedance (b) phase angle.

(b) MULTI-CHOICE PROBLEMS (answers on page 125)

1 An inductance of 10 mH connected across a 100 V, 50 Hz supply has an inductive reactance of (a) $10\pi\Omega$ (b) $1000\pi\Omega$ (c) $\pi\Omega$ (d) πH.

2 When the frequency of an a.c. circuit containing resistance and inductance is increased, the current (a) decreases, (b) increases, (c) stays the same.

3 In problem 2, the phase angle of the circuit, (a) decreases, (b) increases, (c) stays the same.

4 A capacitor of 1 μF is connected to a 50 Hz supply. The capactive reactance is
(a) 50 MΩ (b) $\dfrac{10}{\pi}$ kΩ (c) $\dfrac{\pi}{10^4}\,\Omega$ (d) $\dfrac{10}{\pi}\,\Omega$

5 In a series a.c. circuit the voltage across a pure inductance is 12 V and the voltage across a pure resistance is 5 V. The supply voltage is (a) 13 V, (b) 17 V, (c) 7 V, (d) 2.4 V.

6 Inductive reactance results in a current that
   (a) leads the voltage by 90°
   (b) is in phase with the voltage,
   (c) leads the voltage by π rads,
   (d) lags the voltage by $\dfrac{\pi}{2}$ rads.

7 The impedance of a coil, which has a resistance of X ohms and an inductance of Y henry's, connected across a supply of frequency K Hz is: (a) $2\pi KY$ (b) $X + Y$ (c) $\sqrt{(X^2 + Y^2)}$ (d) $\sqrt{X^2 + (2\pi KY)^2}$

8 In problem 7, the phase angle between the current and the applied voltage is given by: (a) arctan $\dfrac{Y}{X}$ (b) arctan $\dfrac{2\pi KY}{X}$ (c) arctan $\dfrac{X}{2\pi KY}$ (d) tan $\dfrac{2\pi KY}{X}$

9 When a capacitor is connected to an a.c. supply the current (a) leads the voltage by 180°, (b) is in phase with the voltage, (c) leads the voltage by $\pi/2$ rads, (d) lags the voltage by 90°.

10 When the frequency of an a.c. circuit containing resistance and capacitance is decreased the current (a) increases, (b) decreases, (c) stays the same.

(c) CONVENTIONAL PROBLEMS

Inductance
1 Calculate the reactance of a coil of inductance 0.2 H when it is connected to (a) a 50 Hz, (b) a 600 Hz (c) a 40 kHz supply.

[(a) 62.83Ω (b) 754Ω (c) 50.27 kΩ]

2 A coil has a reactance of 120Ω in a circuit with a supply frequency of 4 kHz. Calculate the inductance of the coil.

[4.77 mH]

3 A supply of 240 V, 50 Hz is connected across a pure inductance and the resulting current is 1.2 A. Calculate the inductance of the coil.

[0.637 H]

4 An e.m.f. of 200 V at a frequency of 2 kHz is applied to a coil of pure inductance 50 mH. Determine (a) the reactance of the coil, (b) the current flowing in the coil.

[(a) 628Ω (b) 0.318 A]

5 A 120 mH inductor has a 50 mA, 1 kHz alternating current flowing through it. Find the p.d. across the inductor.

[37.7 V]

6 Determine the impedance of a coil which has a resistance of 12Ω and a reactance of 16Ω.

[20Ω]

7 A coil of inductance 80 mH and resistance 60Ω is connected to a 200 V, 100 Hz supply. Calculate the circuit impedance and the current taken from the supply. Find also the phase angle between the current and the supply voltage.

[78.27Ω; 2.555 A; 39° 57' lagging]

8 An alternating voltage given by $v = 100 \sin 240\,t$ volts is applied across a coil of resistance 32Ω and inductance 100 mH. Determine (a) the circuit impedance, (b) the current flowing, (c) the p.d. across the resistance, (d) the p.d. across the inductance. [(a) 40Ω (b) 1.77 A (c) 56.64 V (d) 42.48 V]

9 A coil takes a current of 5 A from a 20 V d.c. supply. When connected to a 200 V, 50 Hz a.c. supply the current is 25 A. Calculate the (a) resistance, (b) impedance and (c) inductance of the coil. [(a) 4Ω (b) 8Ω (c) 22.06 mH]

10 A resistor and an inductor of negligible resistance are connected in series to an a.c. supply. The p.d. across the resistor is 18 V and the p.d. across the inductor is 24 V. Calculate the supply voltage and the phase angle between voltage and current. [30 V, 53° 8' lagging]

11 A coil of inductance 636.6 mH and negligible resistance is connected in series with a 100Ω resistor to a 240 V, 50 Hz supply. Calculate (a) the inductive reactance of the coil, (b) the impedance of the circuit, (c) the current in the circuit, (d) the p.d. across each component, and (e) the circuit phase angle.

[(a) 200Ω (b) 223.6Ω (c) 1.118 A (d) 223.6 V, 111.8 V (e) 63° 26' lagging]

Capacitance

12 Calculate the capacitive reactance of a capacitor of 20 μF when connected to an a.c. circuit of frequency (a) 20 Hz, (b) 50 Hz, (c) 4 kHz.

[(a) 397.9Ω (b) 15.92Ω (c) 1.989Ω]

13 A capacitor has a reactance of 80Ω when connected to a 50 Hz supply. Calculate the value of its capacitance.

[39.79μF]

14 Calculate the current taken by a 10μF capacitor when connected to a 200 V, 100 Hz supply.

[1.257 A]

15 A capacitor has a capacitive reactance of 400Ω when connected to a 100 V, 25 Hz supply. Determine its capacitance and the current taken from the supply.

[15.92μF; 0.25 A]

16 A voltage of 35 V is applied across a $C-R$ series circuit. If the voltage across the resistor is 21 V, find the voltage across the capacitor.

[28 V]

17 A resistance of 50Ω is connected in series with a capacitance of 20μF. If a supply of 200 V, 100 Hz is connected across the arrangement, find (a) the circuit impedance, (b) the current flowing, (c) the phase angle between voltage and current.

[(a) 93.98Ω (b) 2.128 A (c) 57° 51' leading]

18 A 24.87μF capacitor and a 30Ω resistor are connected in series across a 150 V supply. If the current flowing is 3 A find (a) the frequency of the supply, (b) the p.d. across the resistor and (c) the p.d. across the capacitor.

[(a) 160 Hz; (b) 90 V; (c) 120 V]

19 An alternating voltage $v = 250 \sin 800\,t$ volts is applied across a series circuit containing a 30Ω resistor and 50μF capacitor. Calculate (a) the circuit impedance, (b) the current flowing, (c) the p.d. across the resistor, (d) the p.d. across the capacitor, (e) the phase angle between voltage and current.

[(a) 39.05Ω; (b) 4.526 A; (c) 135.8 V; (d) 113.2 V; (e) 39° 48']

20 A 400Ω resistor is connected in series with a 2358 pF capacitor across a 12 V a.c. supply. Determine the supply frequency if the current flowing in the circuit is 24 mA.

[225 kHz]

4 Series a.c. circuits

A. MAIN POINTS CONCERNED WITH SERIES A.C. CIRCUITS

1 In an a.c. series circuit containing resistance R, inductance L and capacitance C, the applied voltage V is the phasor sum of V_R, V_L and V_C (see *Fig 1*). V_L and V_C are anti-phase and there are three phasor diagrams possible, each depending on the relative values of V_L and V_C.

2 When $X_L > X_C$ (see *Fig 1(b)*):

$$Z = \sqrt{[R^2 + (X_L - X_C)^2]}$$

and $\tan \phi = \dfrac{(X_L - X_C)}{R}$

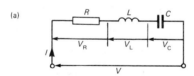

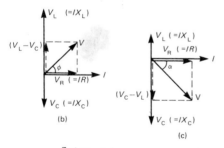

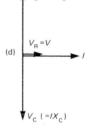

Impedance triangle

Impedance triangle

Fig 1

3   When $X_C > X_L$ (see *Fig 1(c)*):

$$Z = \sqrt{[R^2 + (X_C - X_L)^2]}$$

and $\tan \alpha = \dfrac{(X_C - X_L)}{R}$

4   When $X_L = X_C$ (see *Fig 1(d)*), the applied voltage V and the current I are in phase. This effect is called **series resonance**.

At resonance: (i) $V_L = V_C$

(ii) $Z = R$ (the minimum circuit impedance possible in an L–C–R circuit).

(iii) $I = \dfrac{V}{R}$ (the maximum current possible in an L–C–R circuit).

(iv) $X_L = X_C$, so $2\pi f_0 L = \dfrac{1}{2\pi f_0 C}$,

from which $f_0 = \dfrac{1}{2\pi\sqrt{(LC)}}$ Hz, where f_0 is the

resonance frequency.

5   At resonance, if R is small compared with X_L and X_C, it is possible for V_L and V_C to have voltages many times greater than the supply voltage V (see *Fig 1(d)*).

Voltage magnification at resonance = $\dfrac{\text{voltage across } L \text{ (or } C)}{\text{supply voltage } V}$

(This magnification is termed the Q-**factor**.)

6   (a) For a purely resistive a.c. circuit, the average power dissipated, is given by:

$$P = VI = I^2 R = \dfrac{V^2}{R} \text{ watts } (V \text{ and } I \text{ being r.m.s. values}).$$

See *Fig 2(a)*.

(b) For a purely inductive a.c. circuit, the average power is zero. See *Fig 2(b)*.

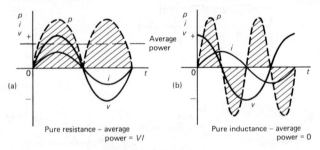

(a) Pure resistance – average power = VI

(b) Pure inductance – average power = 0

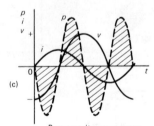

(c) Pure capacitance – average power = 0

Fig 2

(c) For a purely capacitive a.c. circuit, the average power is zero. See *Fig 2(c)*. In *Figs 2(a) – (c)*, the value of power at any instant is given by the product of the voltage and current at that instant, i.e., the instantaneous power, $p = vi$, as shown by the broken line.

7 *Fig 3* shows current and voltage waveforms for an $R-L$ circuit where the current lags the voltage by angle ϕ. The waveform for power (where $p = vi$) is shown by

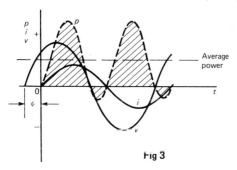

Fig 3

the broken line, and its shape, and hence average power, depends on the value of angle ϕ.

For an $R-L$, $R-C$, or $L-C-R$ series a.c. circuit, the average power P is given by:

$P = VI \cos \phi$ **watts**

or $P = I^2 R$ **watts** (V and I being r.m.s. values).

8 *Fig 4(a)* shows a phasor diagram in which the current I lags the applied voltage V by angle ϕ. The horizontal component of V is $V \cos \phi$ and the vertical component of V is $V \sin \Phi$. If each of the voltage phasors is multiplied by I, *Fig 4(b)* is obtained and is known as the **power triangle**.

Apparent power $S = VI$ voltamperes (VA)
True or active power $P = VI \cos \phi$ watts (W)
Reactive power $Q = VI \sin \phi$ reactive voltamperes (VAr)

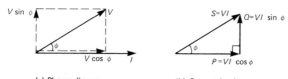

(a) Phasor diagram   (b) Power triangle

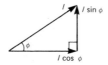

Fig 4   (c) Current triangle

37

9   If each of the phasors of the power triangle of *Fig 4(b)* is divided by *V*, *Fig 4(c)* is obtained and is known as the **current triangle**. The horizontal component of current, $I \cos \phi$, is called the active or the **in-phase component**. The vertical component of current, $I \sin \phi$, is called the **reactive** or the **quadrature component**.

10 Power factor = $\dfrac{\text{True power } P}{\text{Apparent power } S}$

For sinusoidal voltages and currents, power factor = $\dfrac{P}{S} = \dfrac{VI \cos \phi}{VI}$

i.e. p.f. $= \cos \phi = \dfrac{R}{Z}$ (from *Fig 1*)

(The relationships stated in *Paras 8 to 10* are also true when current *I* leads voltage *V*.)

B. WORKED PROBLEMS ON SERIES A.C. CIRCUITS

Problem 1 A coil of resistance 5Ω and inductance 120 mH in series with a 100μF capacitor, is connected to a 300 V, 50 Hz supply. Calculate (a) the current flowing, (b) the phase difference between the supply voltage and current, (c) the voltage across the coil, (d) the voltage across the capacitor.

The circuit diagram is shown in *Fig 5*.

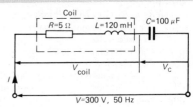

Fig 5

$X_L = 2\pi fL = 2\pi(50)(120 \times 10^{-3}) = 37.70\Omega$

$X_C = \dfrac{1}{2\pi fC} = \dfrac{1}{2\pi(50)(100 \times 10^{-6})} = 31.83\Omega$

Since X_L is greater than X_C the circuit is inductive (see phasor diagram in *Fig 1(b)*).

$X_L - X_C = 37.70 - 31.83 = 5.87\Omega$

Impedance, $Z = \sqrt{[R^2 + (X_L - X_C)^2]} = \sqrt{[(5)^2 + (5.87)^2]} = 7.71\Omega$

(a) Current, $I = \dfrac{V}{Z} = \dfrac{300}{7.71} = \mathbf{38.91\ A}$

(b) Phase angle $\phi = \arctan\left(\dfrac{X_L - X_C}{R}\right) = \arctan\left(\dfrac{5.87}{5}\right) = \mathbf{49°\ 35'}$

(c) Impedance of coil, $Z_{\text{COIL}} = \sqrt{(R^2 + X_L^2)} = \sqrt{[(5)^2 + (37.7)^2]} = 38.03\Omega$

Voltage across coil, $V_{\text{COIL}} = I Z_{\text{COIL}} = (38.91)(38.03) = \mathbf{1480\ V}$

(d) Voltage across capacitor, $V_C = I X_C = (38.91)(31.83) = \mathbf{1239\ V}$

Problem 2 A 25 V, 1 kHz supply is connected across a coil having an inductance of 0.60 mH and resistance 2 Ω. Determine the supply current and its phase angle, and the active and reactive components of the current, showing each on a phasor diagram.

Inductive reactance, $X_L = 2\pi f L = 2\pi(1 \times 10^3)(0.60 \times 10^{-3}) = 3.77\Omega$

Impedance $Z = \sqrt{(R^2 + X_L^2)} = \sqrt{(2^2 + 3.77^2)} = 4.268\Omega$

Supply current $I = \dfrac{V}{Z} = \dfrac{25}{4.268} = \mathbf{5.858\ A}$

Phase angle, $\phi = \arctan \dfrac{X_L}{R}$

$= \arctan \dfrac{3.77}{2} = \mathbf{62°\ 3'\ lagging}$

Active component of current

$= I \cos \phi$, where $\cos \phi = \dfrac{R}{Z}$

$= (5.858)\left(\dfrac{2}{4.268}\right) = \mathbf{2.745\ A}$

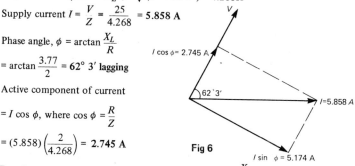

Fig 6

Reactive component of current $= I \sin \phi$, where $\sin \phi = \dfrac{X_L}{Z}$

$$= (5.858)\left(\dfrac{3.77}{4.268}\right) = \mathbf{5.174\ A}$$

The phasor diagram is shown in *Fig 6* where the active (or 'in-phase' component) is 'in-phase' with the supply voltage V.

Problem 3 A coil of resistance 75Ω and inductance 150 mH in series with an 8μF capacitor, is connected to a 500 V, 200 Hz supply. Calculate (a) the current flowing, (b) the phase difference between the supply voltage and current, (c) the voltage across the coil and (d) the voltage across the capacitor. Sketch the phasor diagram.

The circuit diagram is as shown in *Fig 1(a)*.

Inductive reactance, $X_L = 2\pi f L = 2\pi(200)(150 \times 10^{-3}) = 188.5\Omega$

Capacitive reactance, $X_C = \dfrac{1}{2\pi f C} = \dfrac{1}{2\pi(200)(8 \times 10^{-6})} = 99.47\Omega$

Since $X_L > X_C$ the circuit is inductive (see phasor diagram in *Fig 1(b)*).

$X_L - X_C = 188.5 - 99.47 = 89.03\Omega$.

Impedance $Z = \sqrt{[R^2 + (X_L - X_C)^2]} = \sqrt{[(75)^2 + (89.03)^2]} = 116.4\Omega$

(a) Current $I = \dfrac{V}{Z} = \dfrac{500}{116.4} = \mathbf{4.296\ A}$

(b) From *Fig 1(b)*, phase angle $\phi = \arctan\left(\dfrac{X_L - X_C}{R}\right) = \arctan\left(\dfrac{89.03}{75}\right)$

$= \mathbf{49°\ 53'\ lagging.}$

(c) Impedance of coil $Z_{COIL} = \sqrt{(R^2 + X_L^2)} = \sqrt{[75^2 + 188.5^2]} = 202.9\Omega$

Voltage across coil $V_{COIL} = I Z_{COIL} = (4.296)(202.9) = \mathbf{871.7\ V}$

Phase angle of coil, $\theta = \arctan\left(\dfrac{X_L}{R}\right) = \arctan\left(\dfrac{188.5}{75}\right) = \mathbf{68°\ 18'\ lagging}$

(d) Voltage across capacitor, $V_C = I X_C = (4.296)(99.47) = \mathbf{427.3\ V}$

The phasor diagram is shown in *Fig 7*. The supply voltage V is the phasor sum of V_{COIL} and V_C

39

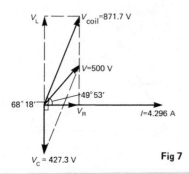

Fig 7

Problem 4 The following three impedances are connected in series across a 40 V, 20 kHz supply: (i) a resistance of 8Ω, (ii) a coil of inductance 130μH and 5Ω resistance and (iii) a 10Ω resistor in series with a 0.25μF capacitor. Calculate (a) the circuit current, (b) the circuit phase angle and (c) the voltage drop across each impedance.

The circuit diagram is shown in *Fig 8(a)*. Since the total circuit resistance is 8 + 5 + 10, or 23Ω, an equivalent circuit diagram may be drawn as *Fig 8(b)*.

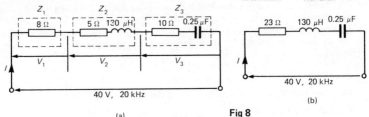

Fig 8

Inductive reactance, $X_L = 2\pi f L = 2\pi(20 \times 10^3)(130 \times 10^{-6}) = 16.34\Omega$

Capacitive reactance, $X_C = \dfrac{1}{2\pi f C} = \dfrac{1}{2\pi(20 \times 10^3)(0.25 \times 10^{-6})} = 31.83\Omega$

As $X_C > X_L$, the circuit is capacitive (see phasor diagram in *Fig 1(c)*).
$X_C - X_L = 31.83 - 16.34 = 15.49\Omega$.

(a) Circuit impedance, $Z = \sqrt{[R^2 + (X_C - X_L)^2]} = \sqrt{[23^2 + 15.49^2]} = 27.73\Omega$

Circuit current, $I = \dfrac{V}{Z} = \dfrac{40}{27.73} = \mathbf{1.442\ A}$

(b) From *Fig 1(c)*, circuit phase angle $\phi = \arctan\left(\dfrac{X_C - X_L}{R}\right) = \arctan\left(\dfrac{15.49}{23}\right)$
$= \mathbf{33°\ 58'\ leading}$

(c) From *Fig 8(a)*, $V_1 = I R_1 = (1.442)(8) = \mathbf{11.54\ V}$
$V_2 = I Z_2 = I\sqrt{(5^2 + 16.34^2)} = (1.442)(17.09) = \mathbf{24.64\ V}$
$V_3 = I Z_3 = I\sqrt{(10^2 + 31.83^2)} = (1.442)(33.36) = \mathbf{48.11\ V}$

The 40 V supply voltage is the phasor sum of V_1, V_2 and V_3.

40

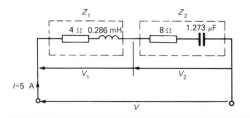

Fig 9

For impedance Z_1 : $R_1 = 4\Omega$ and $X_L = 2\pi fL = 2\pi(5 \times 10^3)(0.286 \times 10^{-3}) = 8.985\Omega$

$$V_1 = I Z_1 = I\sqrt{(R^2 + X_L^2)} = 5\sqrt{(4^2 + 8.985^2)} = 49.18 \text{ V}$$

$$\text{Phase angle } \phi_1 = \arctan\left(\frac{X_L}{R}\right) = \arctan\left(\frac{8.985}{4}\right) = 66° \ 0' \text{ lagging}$$

For impedance Z_2 : $R_2 = 8\Omega$ and $X_C = \frac{1}{2\pi fC} = \frac{1}{2\pi(5 \times 10^3)(1.273 \times 10^{-6})} = 25.0\Omega$

$$V_2 = I Z_2 = I\sqrt{(R^2 + X_C^2)} = 5\sqrt{(8^2 + 25.0^2)} = 131.2 \text{ V}$$

$$\text{Phase angle } \phi_2 = \arctan\left(\frac{X_C}{R}\right) = \arctan\left(\frac{25.0}{8}\right) = 72° \ 15' \text{ leading}$$

The phasor diagram is shown in *Fig 10*. The phasor sum of V_1 and V_2 gives the supply voltage V of **100 V** at a phase angle of **53° 8' leading**. These values may be determined by drawing or by calculation — either by resolving into horizontal and vertical components or by the cosine and sine rules.

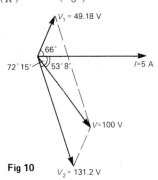

Fig 10

$$\text{Resonance frequency}, f_o = \frac{1}{2\pi\sqrt{(LC)}} \text{ Hz} = \frac{1}{2\pi\sqrt{\left[\left(\frac{125}{10^3}\right)\left(\frac{60}{10^6}\right)\right]}} \text{ Hz}$$

$$= \frac{1}{2\pi\sqrt{\left(\frac{125 \times 6}{10^8}\right)}} = \frac{1}{\frac{2\pi\sqrt{(125)(6)}}{10^4}}$$

$$= \frac{10^4}{2\pi\sqrt{(125)(6)}} = 58.12 \text{ Hz}$$

At resonance, $X_L = X_C$ and impedance $Z = R$

Hence current $I = \dfrac{V}{R} = \dfrac{120}{10} = 12$ A

Problem 7 The current at resonance in a series $L-C-R$ circuit is 100 μA. If the applied voltage is 2 mV at a frequency of 200 kHz, and the circuit inductance is 50 μH, find (a) the circuit resistance, and (b) the circuit capacitance.

(a) $I = 100 \ \mu\text{A} = 100 \times 10^{-6}$ A; $V = 2$ mV $= 2 \times 10^{-3}$ V.

   At resonance, impedance Z = resistance R.

   Hence $R = \dfrac{V}{I} = \dfrac{2 \times 10^{-3}}{100 \times 10^{-6}} = \dfrac{2 \times 10^6}{100 \times 10^3} = 20\Omega$

(b) At resonance $X_L = X_C$

   i.e. $2\pi f L = \dfrac{1}{2\pi f C}$

   Hence capacitance $C = \dfrac{1}{(2\pi f)^2 L} = \dfrac{1}{(2\pi \times 200 \times 10^3)^2 (50 \times 10^{-6})}$ F

   $\qquad = \dfrac{(10^6)(10^6)}{(4\pi)^2(10^{10})(50)} \ \mu\text{F} = 0.0127 \ \mu\text{F}$

   $\qquad\qquad\qquad\qquad\qquad$ or 12.7 nF.

Problem 8 A coil of inductance 80 mH and negligible resistance is connected in series with a capacitance of 0.25 μF and a resistor of resistance 12.5Ω across a 100 V, variable-frequency supply. Determine (a) the resonance frequency, and (b) the current at resonance. How many times greater than the supply voltage is the voltage across the reactances at resonance?

(a) Resonance frequency, $f_o = \dfrac{1}{2\pi\sqrt{\left[\left(\frac{80}{10^3}\right)\left(\frac{0.25}{10^6}\right)\right]}} = \dfrac{1}{2\pi\sqrt{\left(\frac{(8)(0.25)}{10^8}\right)}} = \dfrac{10^4}{2\pi\sqrt{2}}$

$\qquad\qquad\qquad\qquad\qquad\qquad = 1125.4 \text{ Hz} = 1.1254 \text{ kHz}$

(b) Current at resonance, $I = \dfrac{V}{R} = \dfrac{100}{12.5} \qquad = 8$ A

Voltage across inductance at resonance $V_L = I X_L = (I)(2\pi f L)$
$\qquad\qquad\qquad\qquad\qquad\qquad\qquad = (8)(2\pi)(1125.4)(80 \times 10^{-3})$
$\qquad\qquad\qquad\qquad\qquad\qquad\qquad = 4525.5$ V

Also, voltage across capacitor, $V_C = I X_C = \dfrac{I}{2\pi f C} = \dfrac{8}{2\pi(1125.4)(0.25 \times 10^{-6})}$
$\qquad\qquad\qquad\qquad\qquad\qquad\qquad = 4525.5$ V

Voltage magnification at resonance $= \dfrac{V_L}{V}$ (or $\dfrac{V_C}{V}$) $= \dfrac{4525.5}{100} = \mathbf{45.255}$,

i.e. at resonance, the voltages across the reactances are 45.255 times greater than the supply voltage. Hence Q factor of circuit is 45.255.

Problem 9 An instantaneous current, $i = 250 \sin \omega t$ mA flows through a pure resistance of 5 kΩ. Find the power dissipated in the resistor.

Power dissipated, $P = I^2 R$ where I is the r.m.s. value of current.
If $i = 250 \sin \omega t$ mA, then $I_{MAX} = 0.250$ A and r.m.s. current,
$$I = (0.707 \times 0.250) \text{ A.}$$
Hence power $P = (0.707 \times 0.250)^2 (5000)$ $= \mathbf{156.2 \ watts.}$

Problem 10 A series circuit of resistance 60Ω and inductance 75 mH is connected to a 110 V, 60 Hz supply. Calculate the power dissipated.

Inductive reactance, $X_L = 2\pi f L = 2\pi(60)(75 \times 10^{-3}) = 28.27\Omega$
Impedance, $Z = \sqrt{(R^2 + X_L^2)} = \sqrt{[(60)^2 + (28.27)^2]} = 66.33\Omega$
Current, $I = \dfrac{V}{Z} = \dfrac{110}{66.33} = 1.658$ A.

To calculate power dissipation in an a.c. circuit two formulae may be used:
  (i) $P = I^2 R = (1.658)^2 (60) = \mathbf{165 \ W}$

or (ii) $P = VI \cos \phi$ where $\cos \phi = \dfrac{R}{Z} = \dfrac{60}{66.33} = 0.9046$
Hence $P = (110)(1.658)(0.9046) = \mathbf{165 \ W}$

Problem 11 A pure inductance is connected to a 150 V, 50 Hz supply, and the apparent power of the circuit is 300 VA. Find the value of the inductance.

Apparent power $S = V I$

Hence current $\quad I = \dfrac{S}{V} = \dfrac{300}{150} = 2$ A

Inductive reactance, $X_L = \dfrac{V}{I} = \dfrac{150}{2} = 75\Omega$

Since $X_L = 2\pi f L$, inductance $L = \dfrac{X_L}{2\pi f} = \dfrac{75}{2\pi(50)} = \mathbf{0.239 \ H}$

Problem 12 A transformer has a rated output of 200 kVA at a power factor of 0.8. Determine the rated power output and the corresponding reactive power.

$VI = 200$ kVA $= 200 \times 10^3$: p.f. $= 0.8 = \cos \phi$.
Power output, $P = VI \cos \phi = (200 \times 10^3)(0.8) = \mathbf{160 \ kW}$
Reactive power, $Q = VI \sin \phi$
If $\cos \phi = 0.8$, then $\phi = \arccos 0.8 = 36° \ 52'$.
Hence $\sin \phi = \sin 36° \ 52' = 0.6$
Hence reactive power, $Q = (200 \times 10^3)(0.6) = \mathbf{120 \ kVAr}$

Problem 13 A load takes 90 kW at a power factor of 0.5 lagging. Calculate the apparent power and the reactive power.

True power, $P = 90$ kW $= V I \cos \phi$

Power factor, $= 0.5 \quad = \cos \phi$

Apparent power $S = VI = \dfrac{90 \text{ kW}}{0.5} = \mathbf{180 \ kVA}$

Angle $\phi = \arccos 0.5 = 60°$. Hence $\sin \phi = \sin 60° = 0.866$
Hence reactive power, $Q = VI \sin \phi = 180 \times 10^3 \times 0.866 = \mathbf{156 \ kVAr}$

Problem 14 The power taken by an inductive circuit when connected to a 120 V, 50 Hz supply is 400 watts and the current is 8 A. Calculate (a) the resistance, (b) the impedance, (c) the reactance, (d) the power factor, (e) the phase angle between voltage and current.

(a) Power $P = I^2 R$. Hence $R = \dfrac{P}{I^2} = \dfrac{400}{(8)^2} = 6.25\Omega$

(b) Impedance $Z = \dfrac{V}{I} = \dfrac{120}{8} = 15\Omega$

(c) Since $Z = \sqrt{(R^2 + X_L{}^2)}$, then $X_L = \sqrt{(Z^2 - R^2)} = \sqrt{[(15)^2 - (6.25)^2]}$
$= 13.64\Omega$

(d) Power factor $= \dfrac{\text{true power}}{\text{apparent power}} = \dfrac{VI \cos \phi}{VI} = \dfrac{400}{(120)(8)} = 0.4167$

(e) p.f. $\cos \phi = 0.4167$. Hence phase angle $\phi = \arccos 0.4167 = 65° \, 22'$ **lagging.**

Problem 15 A circuit consisting of a resistor in series with a capacitor takes 100 watts at a power factor of 0.5 from a 100 V, 60 Hz supply. Find (a) the current flowing, (b) the phase angle, (c) the resistance, (d) the impedance, (e) the capacitance.

(a) Power factor $= \dfrac{\text{true power}}{\text{apparent power}}$, or $0.5 = \dfrac{100}{100\,I}$

Hence $I = \dfrac{100}{(0.5)(100)} = 2\text{A}$

(b) Power factor $= 0.5 = \cos \phi$. Hence phase angle $\phi = \arccos 0.5 = 60°$ **leading.**

(c) Power $P = I^2 R$. Hence resistance $R = \dfrac{P}{I^2} = \dfrac{100}{(2)^2} = 25\Omega$

(d) Impedance $Z = \dfrac{V}{I} = \dfrac{100}{2} = 50\Omega$

(e) Capacitive reactance, $X_C = \sqrt{(Z^2 - R^2)} = \sqrt{(50^2 - 25^2)} = 43.30\Omega$

$X_C = \dfrac{1}{2\pi f C}$. Hence capacitance $C = \dfrac{1}{2\pi f X_C} = \dfrac{1}{2\pi(60)(43.30)}$ F
$= 61.26 \, \mu\text{F}$

Problem 16 A single-phase motor is connected to a 400 V, 50 Hz supply. The motor develops 15 kW with an efficiency of 80% and a power factor of 0.75 lagging. Determine (a) the input kVA's, (b) the active and reactive components of the current, (c) the reactive voltamperes.

(a) Efficiency $= \dfrac{\text{output power}}{\text{input power}} = \dfrac{\text{output power}}{VI \cos \phi} = \dfrac{\text{output power}}{(VI)(\text{p.f.})}$

Hence $\dfrac{80}{100} = \dfrac{15 \times 10^3}{(VI)(0.75)}$, from which, $VI = \dfrac{15 \times 10^3}{(0.80)(0.75)} = 25\,000$ VA.

Therefore, input kilovoltamperes $= 25$ **kVA.**

(b) Current taken by motor $= \dfrac{\text{input voltamperes}}{\text{voltage}} = \dfrac{VI}{V} = \dfrac{25\,000}{400}$
$= 62.5$ A.

Active or in-phase component of current $= I \cos \phi = (62.5)(0.75)$
$= 46.88$ **A**

Since p.f. = cos ϕ = 0.75 then ϕ = arccos 0.75 = 41.41°.
Hence sin ϕ = sin 41.41° = 0.6614.
Reactive or quadrative component of current = I sin ϕ = (62.5)(0.6614)
$$= 41.34 \text{ A}$$
(c) Reactive voltamperes = $V I$ sin ϕ = (400)(41.34) = **16.54 kV Ar.**

C. FURTHER PROBLEMS ON SERIES A.C. CIRCUITS

(a) SHORT ANSWER PROBLEMS

1 Sketch the three possible phasor diagrams for a series circuit containing resistance, inductance and capacitance.
2 What is series resonance?
3 Derive a formula for resonance frequency f_o, in terms of L and C.
4 State two formulae which may be used to calculate power in an a.c. circuit.
5 Sketch a phasor diagram in which the current leads the applied voltage by angle β. From the 'voltage triangle' derive (a) the power triangle, (b) the current triangle.
6 What is meant by the Q-factor of a circuit?
7 What is meant by (a) the active component of current, (b) the reactive component of current?
8 Define 'power factor'.
9 Define (a) apparent power, (b) reactive power.
10 Show graphically that for a purely inductive or purely capacitive a.c. circuit the average power is zero.

(b) MULTI-CHOICE PROBLEMS (answers on page 125)

1 Which of the following statements is false?
   (a) Impedance is at a minimum at resonance in an a.c. circuit.
   (b) The product of r.m.s. current and voltage gives the apparent power in an a.c. circuit.
   (c) Current is at a maximum at resonance in an a.c. circuit.
   (d) $\dfrac{\text{Apparent power}}{\text{true power}}$ gives power factor.
2 In an $R-L-C$ a.c. circuit, a current of 5 A flows when the supply voltage is 100 V. The phase angle between current and voltage is 60° lagging. Which of the following statements is false?
   (a) The circuit is effectively inductive.
   (b) The apparent power is 500 VA.
   (c) The equivalent circuit reactance is 20Ω.
   (d) The true power is 250 W.
3 The current flowing in a coil is 5 A. The phase angle between the current and the applied voltage of 10 V is 30° lagging. Which of the following statements is true?
   (a) Active power = 50 watts.
   (b) The circuit is effectively capacitive.
   (c) The active component of current is 5 A.
   (d) The reactive component of current is 2.5 A.
4 In an $R-L-C$ a.c. series circuit a current of 2 A flows when the supply voltage is

100 V. The phase angle between current and voltage is 60° leading. Which of the following statements is false?

(a) The circuit is effectively capacitive.

(b) The apparent power is 100 W.

(c) The equivalent circuit impedance is 50Ω.

(d) The power factor is 0.5 leading.

5   A series a.c. circuit comprising a coil of inductance 100 mH and resistance 1Ω and a 10 μF capacitor is connected across a 10 V supply. At resonance the p.d. across the capacitor is (a) 10 kV, (b) 1 kV, (c) 100 V, (d) 10 V.

(c) CONVENTIONAL PROBLEMS

1   A 40 μF capacitor in series with a coil of resistance 8Ω and inductance 80 mH is connected to a 200 V, 100 Hz supply. Calculate (a) the circuit impedance, (b) the current flowing, (c) the phase angle between voltage and current, (d) the voltage across the coil, and (e) the voltage across the capacitor.

[(a) 13.18Ω; (b) 15.17 A; (c) 52° 38′; (d) 772.1 V; (e) 603.6 V]

2   A 40 V, 2.5 kHz supply is connected across a coil having an inductance of 0.40 mH and resistance 3Ω. Determine the supply current and its phase angle and the active and reactive components of the current, showing each on a phasor diagram.

[5.745 A; 64° 29′ lagging; 2.475 A; 5.185 A]

3   A coil of inductance 74 mH and resistance 28Ω in series with a 50 μF capacitor is connected to a 250 V, 50 Hz supply. Calculate (a) the impedance, (b) the current flowing, (c) the phase difference between voltage and current, (d) the voltage across the coil, and (e) the voltage across the capacitor. Sketch the phasor diagram.

[(a) 49.16Ω; (b) 5.085 A; (c) 55° 17′ leading; (d) 185.1 V; (e) 323.7 V]

4   Three impedances are connected in series with a 100 V, 2 kHz supply. The impedances comprise:

  (i)   an inductance of 0.45 mH and 2Ω resistance,

  (ii) an inductance of 570 μH and 5Ω resistance, and

  (iii) a capacitor of 10 μF and 3Ω resistance.

Assuming no mutual inductive effects between the two inductances calculate (a) The circuit impedance, (b) the circuit current, (c) the circuit phase angle, and (d) the voltage across each impedance. Draw the phasor diagram.

[(a) 11.12Ω; (b) 8.99 A; (c) 25° 56′ lagging; (d) 53.92 V; 78.53 V; 76.46 V.]

5   For the circuit shown in *Fig 11*, determine the voltages V_1 and V_2 if the supply frequency is 1 kHz. Draw the phasor diagram and hence determine the supply voltage V and the circuit phase angle.

[V_1 = 26.0 V; V_2 = 67.05 V; V = 50 V; 53° 8′ leading]

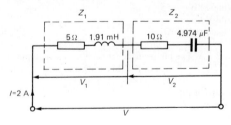

Fig 11

46

6 Find the resonance frequency of a series a.c. circuit consisting of a coil of resistance 10Ω and inductance 50 mH and capacitance $0.05\,\mu F$. Find also the current flowing at resonance if the supply voltage is 100 V.

[3.183 kHz; 10 A]

7 The current at resonance in a series $L-C-R$ circuit is 0.2 mA. If the applied voltage is 250 mV at a frequency of 100 kHz and the circuit capacitance is $0.04\,\mu F$, find the circuit resistance and inductance.

[1.25 kΩ; 63.3 μH]

8 A coil of resistance 25Ω and inductance 100 mH is connected in series with a capacitance of $0.12\,\mu F$ across a 200 V, variable frequency supply. Calculate (a) the resonance frequency, (b) the current at resonance, (c) the factor by which the voltage across the reactance is greater than the supply voltage.

[(a) 1.453 kHz; (b) 8 A; (c) 36.52]

9 A coil of 0.5 H inductance and 8 Ω resistance is connected in series with a capacitor across a 200 V, 50 Hz supply. If the current is in phase with the supply voltage, determine the value of the capacitor and the p.d. across its terminals.

[20.26 μF; 3 928 kV]

10 Calculate the inductance which must be connected in series with a 1000 pF capacitor to give a resonance frequency of 400 kHz.

[0.158 mH]

11 A voltage $v = 200 \sin \omega t$ volts is applied across a pure resistance of 1.5 kΩ. Find the power dissipated in the resistor.

[13.33 W]

12 A 50 μF capacitor is connected to a 100 V, 200 Hz supply. Determine the true power and the apparent power.

[0; 628.3 VA]

13 A motor takes a current of 10 A when supplied from a 250 V a.c. supply. Assuming a power factor of 0.75 lagging, find the power consumed. Find also the cost of running the motor for 1 week continuously if 1 kWh of electricity costs 4.20p.

[1875 W; £13.23]

14 A transformer has a rated output of 100 kVA at a power factor of 0.6. Determine the rated power output and the corresponding reactive power.

[60 kW; 80 kVAr]

15 A substation is supplying 200 kVA and 150 kVAr. Calculate the corresponding power and power factor.

[132 kW; 0.66]

16 A load takes 50 kW at a power factor of 0.8 lagging. Calculate the apparent power and the reactive power.

[62.5 kVA; 37.5 kVAr]

17 A coil of resistance 400Ω and inductance 0.20 H is connected to a 75 V, 400 Hz supply. Calculate the power dissipated in the coil.

[5.452 W]

18 An 80Ω resistor and a 6 μF capacitor are connected in series across a 150 V, 200 Hz supply. Calculate (a) the circuit impedance, (b) the current flowing, (c) the power dissipated in the circuit.

[(a) 154.9Ω; (b) 0.968 A; (c) 75 W]

19 The power taken by a series circuit containing resistance and inductance is 240 watts when connected to a 200 V, 50 Hz supply. If the current flowing is 2 A, find the values of the resistance and inductance.

[60Ω; 255 mH]

20 The power taken by a $C-R$ series circuit, when connected to a 105 V, 2.5 kHz supply, is 0.9 kW and the current is 15 A. Calculate (a) the resistance, (b) the impedance, (c) the reactance, (d) the capacitance, (e) the power factor, (f) the phase angle between voltage and current.

[(a) 4Ω; (b) 7Ω; (c) 5.745Ω; (d) 11.08 μF; (e) 0.571; (f) 55°9′ leading]

21 A circuit consisting of a resistor in series with an inductance takes 210 W at a power factor of 0.6 from a 50 V, 100 Hz supply. Find (a) the current flowing, (b) the circuit phase angle, (c) the resistance, (d) the impedance, (e) the inductance.

[(a) 7A; (b) 53° 8′ lagging; (c) 4.286Ω; (d) 7.143Ω; (e) 9.095 mH]

22 A 200 V, 60 Hz supply is applied to a capacitive circuit. The current flowing is 2 A and the power dissipated is 150 watts. Calculate the values of the resistance and capacitance.

[37.5Ω; 28.61 μF]

23 A single-phase motor is connected to a 415 V, 50 Hz supply. The motor develops 12 kW at an efficiency of 75% and a power factor of 0.8 lagging. Determine (a) the input kV A, (b) the active and reactive components of the current, (c) the reactive voltamperes.

[(a) 20 kV A; (b) 38.55 A; 28.91 A, (c) 12 kV Ar]

5 Parallel a.c. circuits

A. FORMULAE AND DEFINITIONS CONCERNED WITH SINGLE-PHASE PARALLEL A.C. CIRCUITS

1 In parallel circuits, such as those shown in *Fig 1*, the voltage is common to each branch of the network and is thus taken as the reference phasor when drawing phasor diagrams.

2 *R-L* **parallel circuit.** In the two branch parallel circuit containing resistance R and inductance L shown in *Fig 1(a)*, the current flowing in the resistance, I_R, is in-phase with the supply voltage V and the current flowing in the inductance, I_L, lags the supply voltage by 90°. The supply current I is the phasor sum of I_R and I_L and thus the current I lags the applied voltage V by an angle lying between 0° and 90° (depending on the values of I_R and I_L), shown as angle ϕ in the phasor diagram.

From the phasor diagram: $\quad I = \sqrt{(I_R{}^2 + I_L{}^2)}$, (by Pythagoras' theorem)

where $\qquad\qquad\qquad I_R = \dfrac{V}{R}$ and $I_L = \dfrac{V}{X_L}$

$$\tan \phi = \frac{I_L}{I_R}\ ,\ \sin \phi = \frac{I_L}{I}\ \text{and}\ \cos \phi = \frac{I_R}{I}$$

(by trigonometric ratios)

Circuit impedance, $Z = \dfrac{V}{I}$

3 *R-C* **parallel circuit.** In the two branch parallel circuit containing resistance R and capacitance C shown in *Fig 1(b)*, I_R is in-phase with the supply voltage V and the current flowing in the capacitor, I_C, leads V by 90°. The supply current I is the phasor sum of I_R and I_C and thus the current I leads the applied voltage V by an angle lying between 0° and 90° (depending on the values of I_R and I_C), shown as angle α in the phasor diagram.

From the phasor diagram: $\quad I = \sqrt{(I_R{}^2 + I_C{}^2)}$, (by Pythagoras' theorem)

where $\qquad\qquad\qquad I_R = \dfrac{V}{R}$ and $I_C = \dfrac{V}{X_C}$

$\tan \alpha = \dfrac{I_C}{I_R}\ ,\ \sin \alpha = \dfrac{I_C}{I}\ \text{and}\ \cos \alpha = \dfrac{I_R}{I}$ (by trigonometric ratios)

Circuit impedance $Z = \dfrac{V}{I}$

CIRCUIT DIAGRAM   PHASOR DIAGRAM

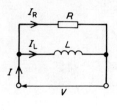 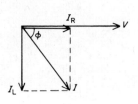

(a)

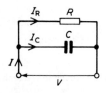 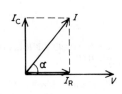

(b)

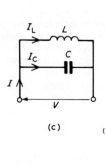 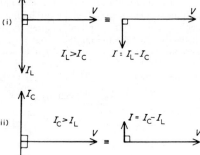

Fig 1

4   *L-C* **parallel circuit.** In the two branch parallel circuit containing inductance L and capacitance C shown in *Fig 1(c)*, I_L lags V by $90°$ and I_C leads V by $90°$. Theoretically there are three phasor diagrams possible—each depending on the relative values of I_L and I_C:

(i) $I_L > I_C$ (giving a supply current, $I = I_L - I_C$ lagging V by $90°$)
(ii) $I_C > I_L$ (giving a supply current, $I = I_C - I_L$ leading V by $90°$)
(iii) $I_L = I_C$ (giving a supply current, $I = 0$).

The latter condition is not possible in practice due to circuit resistance inevitably being present (as in the circuit described in para. 5).

For the *L-C* parallel circuit, $I_L = \dfrac{V}{X_L}$, $I_C = \dfrac{V}{X_C}$

I = phasor difference between I_L and I_C, and $Z = \dfrac{V}{I}$.

50

5   *LR-C* **parallel circuit.** In the two branch circuit containing capacitance C in parallel with inductance L and resistance R in series (such as a coil) shown in *Fig 2(a)*, the phasor diagram for the *LR* branch alone is shown in *Fig 2(b)* and the phasor diagram for the *C* branch is shown alone in *Fig 2(c)*. Rotating each and superimposing on one another gives the complete phasor diagram shown in *Fig 2(d)*.

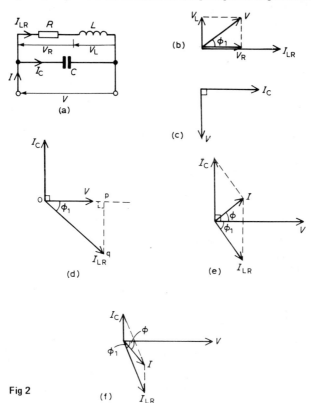

Fig 2

6   The current I_{LR} of *Fig 2(d)* may be resolved into horizontal and vertical components. The horizontal component, shown as op is $I_{LR} \cos \phi_1$, and the vertical component, shown as pq is $I_{LR} \sin \phi_1$. There are three possible conditions for this circuit:

(i) $I_C > I_{LR} \sin \phi_1$ (giving a supply current I leading V by angle ϕ—as shown in *Fig 2(e)*).

(ii) $I_{LR} \sin \phi > I_C$ (giving I lagging V by angle ϕ—as shown in *Fig 2(f)*).

(iii) $I_C = I_{LR} \sin \phi_1$ (this is called parallel resonance, see para. 10).

7   There are two methods of finding the phasor sum of currents I_{LR} and I_C in *Figs 2(e) and (f)*. These are: (i) by a scaled phasor diagram, or (ii) by resolving each current into their 'in-phase' (i.e. horizontal) and 'quadrature' (i.e. vertical) components (see chapter 4, para. 9).

51

8 With reference to the phasor diagrams of *Fig 2*:

Impedance of LR branch, $Z_{LR} = \sqrt{(R^2 + X_L^2)}$

Current, $\qquad I_{LR} = \dfrac{V}{Z_{LR}}$ and $I_C = \dfrac{V}{X_C}$

Supply current I = phasor sum of I_{LR} and I_C (by drawing)

$\qquad\qquad = \sqrt{\{(I_{LR}\cos\phi_1)^2 + (I_{LR}\sin\phi_1 \sim I_C)^2\}}$ (by calculation)

where $\sim$ means 'the difference between'.

Circuit impedance $Z = \dfrac{V}{I}$

$\tan\phi_1 = \dfrac{V_L}{V_R} = \dfrac{X_L}{R}$, $\sin\phi_1 = \dfrac{X_L}{Z_{LR}}$ and $\cos\phi_1 = \dfrac{R}{Z_{LR}}$

$\tan\phi = \dfrac{I_{LR}\sin\phi_1 \sim I_C}{I_{LR}\cos\phi_1}$ and $\cos\phi = \dfrac{I_{LR}\cos\phi_1}{I}$

9 For any parallel a.c. circuit:

True or active power, $\quad P = VI\cos\phi$ watts (W)

$\qquad\qquad$ or $\quad P = I_R^2 R$ watts

Apparent power, $\qquad S = VI$ voltamperes (VA)

Reactive power, $\qquad Q = VI\sin\phi$ reactive voltamperes (VAr)

Power factor $\quad = \dfrac{\text{true power}}{\text{apparent power}} = \dfrac{P}{S} = \cos\phi$

(These formulae are the same as for series a.c. circuits.)

(see *Problems 1 to 7*)

10 (i) **Resonance** occurs in the two branch circuit containing capacitance C in parallel with inductance L and resistance R in series (see *Fig 2(a)*) when the quadrature (i.e. vertical) component of current I_{LR} is equal to I_C. At this condition the supply current I is in-phase with the supply voltage V.

(ii) When the quadrature component of I_{LR} is equal to I_C then:

$I_C = I_{LR}\sin\phi$ (see *Fig 3*)

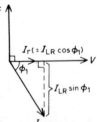

Hence $\dfrac{V}{X_C} = \left(\dfrac{V}{Z_{LR}}\right)\left(\dfrac{X_L}{Z_{LR}}\right)$, (from para. 8)

from which, $Z_{LR}^2 = X_C X_L = (2\pi f_r L)\left(\dfrac{1}{2\pi f_r C}\right) = \dfrac{L}{C}$ (1)

Hence $[\sqrt{(R^2 + X_L^2)}]^2 = \dfrac{L}{C}$ and $R^2 + X_L^2 = \dfrac{L}{C}$

$(2\pi f_r L)^2 = \dfrac{L}{C} - R^2$

Fig 3

$f_r = \dfrac{1}{2\pi L}\sqrt{\left(\dfrac{L}{C} - R^2\right)} = \dfrac{1}{2\pi}\sqrt{\left(\dfrac{L}{L^2 C} - \dfrac{R^2}{L^2}\right)}$

i.e. parallel resonant frequency, $f_r = \dfrac{1}{2\pi}\sqrt{\left(\dfrac{1}{LC} - \dfrac{R^2}{L^2}\right)}$ Hz

(When R is negligible, then $f_r = \dfrac{1}{2\pi\sqrt{(LC)}}$, which is the same as for series resonance.)

(iii) Current at resonance,

$$I_r = I_{LR} \cos \phi_1 \quad \text{(from } \textit{Fig 3}\text{)}$$

$$= \left(\frac{V}{Z_{LR}}\right) \left(\frac{R}{Z_{LR}}\right) \quad \text{(from para 8)}$$

$$= \frac{VR}{Z_{LR}^2}$$

However from equation (1), $Z_{LR}^2 = \dfrac{L}{C}$

Hence $I_r = \dfrac{VR}{\dfrac{L}{C}} = \dfrac{VRC}{L}$ (2)

The current is at a minimum at resonance.

(iv) Since the current at resonance is in-phase with the voltage the impedance of the circuit acts as a resistance. This resistance is known as the **dynamic resistance**, R_D (or sometimes, the dynamic impedance).

From equation (2), impedance at resonance $= \dfrac{V}{I_r} = \dfrac{L}{RC}$

i.e. dynamic resistance, $\quad R_D = \dfrac{L}{RC}$ ohms

(v) The parallel resonant circuit is often described as a **rejector** circuit since it presents its maximum impedance at the resonant frequency and the resultant current is a minimum.

11 Currents higher than the supply current can circulate within the parallel branches of a parallel resonant circuit, the current leaving the capacitor and establishing the magnetic field of the inductor, this then collapsing and recharging the capacitor, and so on. The **Q-factor** of a parallel resonant circuit is the ratio of the current circulating in the parallel branches of the circuit to the supply current, i.e. the current magnification.

$$\text{Q-factor at resonance} = \text{current magnification} = \frac{\text{circulating current}}{\text{supply current}}$$

$$= \frac{I_C}{I_r} = \frac{I_{LR} \sin \phi_1}{I_r}$$

$$= \frac{I_{LR} \sin \phi_1}{I_{LR} \cos \phi_1} = \frac{\sin \phi_1}{\cos \phi_1} = \tan \phi_1 = \frac{X_L}{R}$$

i.e. **Q-factor at resonance** $= \dfrac{2\pi f_r L}{R}$ (which is the same as for a series circuit)

(see *Problems 8 to 10*).

Note that in a parallel circuit the Q-factor is a measure of current magnification, whereas in a series circuit it is a measure of voltage magnification.

12 At mains frequencies the Q-factor of a parallel circuit is usually low, typically less than 10, but in radio-frequency circuits the Q-factor can be very high.

13 For a particular power supplied, a high power factor reduces the current flowing in a supply system and therefore reduces the cost of cables, switch-gear, transformers and generators. Supply authorities use tariffs which encourage electricity consumers to operate at a reasonably high power factor.

14 Industrial loads such as a.c. motors are essentially inductive $(R-L)$ and may have a low power factor. One method of improving (or correcting) the power factor of an inductive load is to connect a static capacitor C in parallel with the load (see

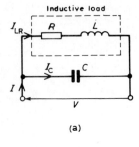

Inductive load

(a)

Fig 4

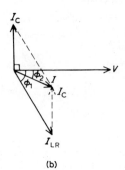

(b)

Fig 4(a)). The supply current is reduced from I_{LR} to I, the phasor sum of I_{LR} and I_C, and the circuit power factor improves from $\cos \phi_1$ to $\cos \phi_2$ (see *Fig 4(b)*).

Another example where capacitors are connected directly across a load occurs in fluorescent lighting, where manufacturers include a power-factor correction capacitor inside the fitting.

15 Where a factory possesses a large number of a.c. motors it may be uneconomical to place a capacitor across the terminals of each motor. As the power factor of an individual motor may vary with load, the capacitor may result in overcorrection at certain loads and even produce a voltage surge that may have a damaging effect on the motor. Many factories have automatic power-factor correction plant situated in their substations, capacitors being switched in or out to maintain the system power-factor between certain predetermined limits. (see *Problems 11 to 14.*)

B. WORKED PROBLEMS ON PARALLEL A.C. CIRCUITS

Problem 1 A 20 Ω resistor is connected in parallel with an inductance of 2.387 mH across a 60 V, 1 kHz supply. Calculate (a) the current in each branch; (b) the supply current; (c) the circuit phase angle; (d) the circuit impedance; and (e) the power consumed.

The circuit and phasor diagrams are as shown in *Fig 1(a)*.

(a) Current flowing in the resistor, $I_R = \dfrac{V}{R} = \dfrac{60}{20} = 3 \text{ A}$

Current flowing in the inductance, $I_L = \dfrac{V}{X_L} = \dfrac{V}{2\pi fL}$

$$= \dfrac{60}{2\pi(1000)(2.387 \times 10^{-3})} = 4 \text{ A}$$

(b) From the phasor diagram, supply current, $I = \sqrt{(I_R{}^2 + I_L{}^2)} = \sqrt{(3^2 + 4^2)} = \mathbf{5 \text{ A}}$

(c) Circuit phase angle $\phi = \arctan \dfrac{I_L}{I_R} = \arctan \dfrac{4}{3} = 53° \, 8' \textbf{ lagging}$

(d) Circuit impedance, $Z = \dfrac{V}{I} = \dfrac{60}{5} = 12 \, \Omega$

(e) Power consumed $P = VI \cos \phi = (60)(5)(\cos 53°\ 8') = \textbf{180 W}$
(Alternatively, power consumed $P = I_R{}^2 R = (3)^2(20) = \textbf{180 W}$)

The circuit and phasor diagrams are as shown in *Fig 1(b)*.

(a) Current in resistor $\quad I_R = \dfrac{V}{R} = \dfrac{240}{80} = \textbf{3 A}$

$\quad$ Current in capacitor $\quad I_C = \dfrac{V}{X_C} = \dfrac{V}{\dfrac{1}{2\pi fC}} = 2\pi fCV = 2\pi(50)(30 \times 10^{-6})(240)$

$$= \textbf{2.262 A}$$

(b) Supply current $\quad I = \sqrt{(I_R{}^2 + I_C{}^2)} = \sqrt{(3^2 + 2.262^2)} = \textbf{3.757 A}$

(c) Circuit phase angle, $\alpha = \arctan \dfrac{I_C}{I_R} = \arctan \dfrac{2.262}{3} = \textbf{37°\ 1' leading}$

(d) Circuit impedance $Z = \dfrac{V}{I} = \dfrac{240}{3.757} = \textbf{63.88 }\Omega$

(e) True or active power dissipated $P = VI \cos \alpha = (240)(3.757) \cos 37°\ 1'$
$$= \textbf{720 W}$$
$\quad$ (Alternatively, true power $P = I_R{}^2 R = (3)^2(80) = \textbf{720 W}$)

(f) Apparent power, $S = VI = (240)(3.757) = \textbf{901.7 V A}$

The circuit diagram is shown in *Fig 5(a)*.
Power factor = $\cos \phi = 0.6$ leading. Hence $\phi = \arccos 0.6 = 53°\ 8'$ leading.
From the phasor diagram shown in *Fig 5(b)*, $I_R = I \cos 53°\ 8' = (2)(0.6) = 1.2$ A
$\qquad\qquad\qquad$ and $\quad I_C = I \sin 53°\ 8' = (2)(0.8) = 1.6$ A
(Alternatively, I_R and I_C can be measured from a scaled phasor diagram.)

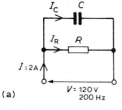 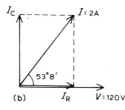

(a) $\qquad V = 120$ V $\quad$ 200 Hz $\qquad$ **Fig 5** $\qquad$ (b)

From the circuit diagram, $I_R = \dfrac{V}{R}$ from which $R = \dfrac{V}{I_R} = \dfrac{120}{1.2} = \textbf{100 }\Omega$

$\qquad$ and $\quad I_C = \dfrac{V}{X_L} = 2\pi fCV$, from which $C = \dfrac{I_C}{2\pi fV}$

$$= \dfrac{1.6}{2\pi(200)(120)}$$

$$= \textbf{10.61 }\mu\textbf{F}$$

55

Problem 4 A pure inductance of 120 mH is connected in parallel with a 25 μF capacitor and the network is connected to a 100 V, 50 Hz supply. Determine (a) the branch currents; (b) the supply current and its phase angle; (c) the circuit impedance; and (d) the power consumed.

The circuit and phasor diagrams are as shown in *Fig 1(c)*.

(a) Inductive reactance $\quad X_L = 2\pi fL = 2\pi(50)(120 \times 10^{-3}) = 37.70\ \Omega$

Capacitive reactance $\quad X_C = \dfrac{1}{2\pi fC} = \dfrac{1}{2\pi(50)(25 \times 10^{-6})} = 127.3\ \Omega$

Current flowing in inductance $I_L = \dfrac{V}{X_L} = \dfrac{100}{37.70} = 2.653$ A

Current flowing in capacitor $\quad I_C = \dfrac{V}{X_C} = \dfrac{100}{127.3} = 0.786$ A

(b) I_L and I_C are anti-phase. Hence supply current $I = I_L - I_C = 2.653 - 0.786$
$\qquad\qquad\qquad\qquad\qquad\qquad\qquad\qquad\qquad\qquad\quad = 1.867$ A

The current I lags the supply voltage V by 90° (see *Fig 1(c i)*).

(c) Circuit impedance $Z = \dfrac{V}{I} = \dfrac{100}{1.867} = 53.56\ \Omega$

(d) Power consumed $P = VI \cos\phi = (100)(1.867)(\cos 90°) = 0$ W

Problem 5 Repeat Problem 4 for the condition when the frequency is changed to 150 Hz

(a) Inductive reactance, $X_L = 2\pi(150)(120 \times 10^{-3}) = 113.1\ \Omega$

Capacitive reactance, $X_C = \dfrac{1}{2\pi(150)(25 \times 10^{-6})} = 42.44\ \Omega$

Current flowing in inductance, $\quad I_L = \dfrac{V}{X_L} = \dfrac{100}{113.1} = 0.884$ A

Current flowing in capacitor, $\quad I_C = \dfrac{V}{X_C} = \dfrac{100}{42.44} = 2.356$ A

(b) Supply current, $I = I_C - I_L = 2.356 - 0.884 = 1.472$ A leading V by 90°
$\qquad\qquad\qquad\qquad\qquad\qquad\qquad\qquad\qquad\qquad$ (see *Fig 1(c ii)*).

(c) Circuit impedance, $Z = \dfrac{V}{I} = \dfrac{100}{1.472} = 67.93\ \Omega$

(d) Power consumed, $P = VI \cos\phi = 0$ W (since $\phi = 90°$)

From Problems 4 and 5:

(i) When $X_L < X_C$ then $I_L > I_C$ and I lags V by 90°.

(ii) When $X_L > X_C$ then $I_L < I_C$ and I leads V by 90°.

(iii) In a parallel circuit containing no resistance the power consumed is zero.

Problem 6 A coil of inductance 159.2 mH and resistance 40 Ω is connected in parallel with a 30 μF capacitor across a 240 V, 50 Hz supply. Calculate (a) the current in the coil and its phase angle, (b) the current in the capacitor and its phase angle, (c) the supply current and its phase angle, (d) the circuit impedance, (e) the power consumed, (f) the apparent power and (g) the reactive power. Draw the phasor diagram.

The circuit diagram is shown in *Fig 6(a)*.

(a) For the coil, inductive reactance $X_L = 2\pi f L = 2\pi(50)(159.2 \times 10^{-3}) - 50\ \Omega$

Impedance $Z_1 = \sqrt{(R^2 + X_L^2)} = \sqrt{(40^2 + 50^2)} = 64.03\ \Omega$

Current in coil, $I_{LR} = \dfrac{V}{Z_1} = \dfrac{240}{64.03} = 3.748$ A

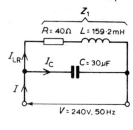

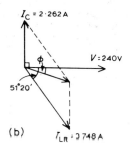

(d) **Fig 6** (b)

Branch phase angle $\phi_1 = \arctan \dfrac{X_L}{R} = \arctan \dfrac{50}{40} = \arctan 1.25$
$= 51°\ 20'$ lagging (see phasor diagram in *Fig 6(b)*)

(b) Capacitive reactance, $X_C = \dfrac{1}{2\pi f C} = \dfrac{1}{2\pi(50)(30 \times 10^{-6})} = 106.1\ \Omega$

Current in capacitor, $I_C = \dfrac{V}{X_C} = \dfrac{240}{106.1} = 2.262$ A **leading the supply voltage by 90°** (see phasor diagram of *Fig 6(b)*).

(c) The supply current I is the phasor sum of I_{LR} and I_C. This may be obtained by drawing the phasor diagram to scale and measuring the current I and its phase angle relative to V. (Current I will always be the diagonal of the parallelogram formed as in *Fig 6(b)*). Alternatively the currents I_{LR} and I_C may be resolved into their horizontal (or 'in-phase') and vertical (or 'quadrature') components. The horizontal component of I_{LR} is

$I_{LR} \cos(51°\ 20') = 3.748 \cos 51°\ 20' = 2.342$ A

The horizontal component of I_C is $I_C \cos 90° = 0$.

Thus the total horizontal component, $I_H = \mathbf{2.342}$ A

The vertical component of $I_{LR} = -I_{LR} \sin(51°\ 20') = -3.748 \sin 51°\ 20'$
$= -2.926$ A

The vertical component of $I_C = I_C \sin 90° = 2.262 \sin 90° = 2.262$ A
Thus the total vertical component, $I_V = -2.926 + 2.262 = \mathbf{-0.664}$ A
I_H and I_V are shown in *Fig 7*, from which,

$I = \sqrt{[(2.342)^2 + (-0.664)^2]} = 2.434$ A

Angle $\phi = \arctan \dfrac{0.664}{2.342} = 15°\ 50'$ lagging.

Hence the supply current $I = 2.434$ A lagging V by $15°\ 50'$.

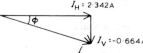

Fig 7

(d) Circuit impedance $Z = \dfrac{V}{I} = \dfrac{240}{2.434} = 98.60\ \Omega$

(e) Power consumed $\quad P = VI \cos \phi = (240)(2.434) \cos 15° 50' = \textbf{562 W}$
(Alternatively $P = I_R{}^2 R = I_{LR}{}^2 R$ (in this case) $= (3.748)^2 (40) = \textbf{562 W}$)
(f) Apparent power $\quad S = VI = (240)(2.434) = \textbf{584.2 V A}$
(g) Reactive power $\quad Q = VI \sin \phi = (240)(2.434)(\sin 15° 50') = \textbf{159.4 V Ar}$

Problem 7 A coil of inductance 0.12 H and resistance 3 kΩ is connected in parallel with a 0.02 μF capacitor and is supplied at 40 V at a frequency of 5 kHz. Determine (a) the current in the coil, and (b) the current in the capacitor. (c) Draw to scale the phasor diagram and measure the supply current and its phase angle. Calculate (d) the circuit impedance and (e) the power consumed.

The circuit diagram is shown in *Fig 8(a)*.
(a) Inductive reactance, $X_L = 2\pi f L = 2\pi(5000)(0.12) = 3770 \ \Omega$
   Impedance of coil, $Z_1 = \sqrt{(R^2 + X_L{}^2)} = \sqrt{[(3000)^2 + (3770)^2]} = 4818 \ \Omega$

   Current in coil, $\quad I_{LR} = \dfrac{V}{Z_1} = \dfrac{40}{4818} = \textbf{8.30 mA}$

   Branch phase angle $\phi = \arctan \dfrac{X_L}{R} = \arctan \dfrac{3770}{3000} = \textbf{51.5° lagging.}$

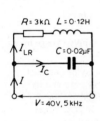

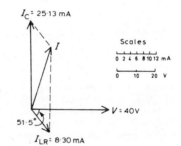

Fig 8

(b) Capacitive reactance, $X_L = \dfrac{1}{2\pi f C} = \dfrac{1}{2\pi(5000)(0.02 \times 10^{-6})} = 1592 \ \Omega$

   Capacitor current, $I_C = \dfrac{V}{X_C} = \dfrac{40}{1592} = \textbf{25.13 mA leading V by 90°.}$

(c) Currents I_{LR} and I_C are shown in the phasor diagram of *Fig 8(b)*. The parallelogram is completed as shown and the supply current is given by the diagonal of the parallelogram. The current I is measured as **19.3 mA** leading voltage V by **74.5°**
   (By calculation, $I = \sqrt{[(I_{LR} \cos 51.5°)^2 + (I_C - I_{LR} \sin 51.5°)^2]} = 19.34 \ A$

   and $\phi = \arctan \left(\dfrac{I_C - I_{LR} \sin 51.5°}{I_{LR} \cos 51.5°} \right) = 74.5°$)

(d) Circuit impedance, $Z = \dfrac{V}{I} = \dfrac{40}{19.3 \times 10^{-3}} = \textbf{2.073 k}\Omega$

(e) Power consumed, $P = VI \cos \phi = (40)(19.3 \times 10^{-3})(\cos 74.5°) = \textbf{206.3 mW}$
   (Alternatively, $P = I_R{}^2 R = I_{LR}{}^2 R = (8.30 \times 10^{-3})^2 (3000) = 206.7 \ mW$)

58

Problem 8 A pure inductance of 150 mH is connected in parallel with a 40 μF capacitor across a 50 V, variable frequency supply. Determine (a) the resonant frequency of the circuit and (b) the current circulating in the capacitor and inductance at resonance.

The circuit diagram is shown in *Fig 9*.

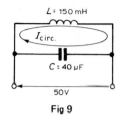

Fig 9

(a) Parallel resonant frequency, $f_r = \dfrac{1}{2\pi}\sqrt{\left(\dfrac{1}{LC} - \dfrac{R^2}{L^2}\right)}$

However, resistance $R = 0$. Hence,

$$f_r = \frac{1}{2\pi}\sqrt{\left(\frac{1}{LC}\right)} = \frac{1}{2\pi}\sqrt{\left(\frac{1}{(150 \times 10^{-3})(40 \times 10^{-6})}\right)}$$

$$= \frac{1}{2\pi}\sqrt{\left(\frac{10^7}{(15)(4)}\right)} = \frac{10^3}{2\pi}\sqrt{\left(\frac{1}{6}\right)} = \textbf{64.97 Hz.}$$

(b) Current circulating in L and C at resonance,

$$I_{CIRC} = \frac{V}{X_C} = \frac{V}{\dfrac{1}{2\pi f_r C}} = 2\pi f_r C V$$

Hence $I_{CIRC} = 2\pi(64.97)(40 \times 10^{-6})(50) = \textbf{0.816 A}$

(Alternatively, $I_{CIRC} = \dfrac{V}{X_L} = \dfrac{V}{2\pi f_r L} = \dfrac{50}{2\pi(64.97)(0.15)} = 0.817$ A)

Problem 9 A coil of inductance 0.20 H and resistance 60 Ω is connected in parallel with a 20 μF capacitor across a 20 V, variable frequency supply. Calculate (a) the resonant frequency; (b) the dynamic resistance; (c) the current at resonance and (d) the circuit Q-factor at resonance.

(a) Parallel resonant frequency, $f_r = \dfrac{1}{2\pi}\sqrt{\left(\dfrac{1}{LC} - \dfrac{R^2}{L^2}\right)}$

$$= \frac{1}{2\pi}\sqrt{\left(\frac{1}{(0.20)(20 \times 10^{-6})} - \frac{(60)^2}{(0.2)^2}\right)}$$

$$= \frac{1}{2\pi}\sqrt{(250\,000 - 90\,000)} = \frac{1}{2\pi}\sqrt{(160\,000)}$$

$$= \frac{1}{2\pi}(400)$$

$$= \textbf{63.66 Hz}$$

(b) Dynamic resistance, $R_D = \dfrac{L}{RC} = \dfrac{0.2}{(60)(20 \times 10^{-6})} = \textbf{166.7 } \Omega$

(c) Current at resonance, $I_r = \dfrac{V}{R_D} = \dfrac{20}{166.7} = \textbf{0.12 A}$

(d) Circuit Q-factor at resonance $= \dfrac{2\pi f_r L}{R} = \dfrac{2\pi(63.66)(0.2)}{60} = \textbf{1.33}$

Alternatively, Q-factor at resonance = current magnification (for a parallel circuit) $= I_c/I_r$

$$I_c = \frac{V}{X_c} = \frac{V}{\frac{1}{2\pi f_r C}} = 2\pi f_r CV = 2\pi(63.66)(20 \times 10^{-6})(20) = 0.16 \text{ A}$$

Hence Q-factor $= \dfrac{I_c}{I_r} = \dfrac{0.16}{0.12} = 1.33$, as obtained above.

Problem 10 A coil of inductance 100 mH and resistance 800 Ω is connected in parallel with a variable capacitor across a 12 V, 5 kHz supply. Determine for the condition when the supply current is a minimum: (a) the capacitance of the capacitor, (b) the dynamic resistance, (c) the supply current, and (d) the Q-factor.

(a) The supply current is a minimum when the parallel circuit is at resonance.

Resonant frequency, $f_r = \dfrac{1}{2\pi}\sqrt{\left(\dfrac{1}{LC} - \dfrac{R^2}{L^2}\right)}$

Transposing for C gives: $(2\pi f_r)^2 = \dfrac{1}{LC} - \dfrac{R^2}{L^2}$

$$(2\pi f_r)^2 + \frac{R^2}{L^2} = \frac{1}{LC}$$

$$C = \frac{1}{L\left\{(2\pi f_r)^2 + \dfrac{R^2}{L^2}\right\}}$$

When $L = 100$ mH, $R = 800 \Omega$ and $f_r = 5000$ Hz,

$$C = \frac{1}{100 \times 10^{-3}\left\{(2\pi 5000)^2 + \dfrac{800^2}{(100 \times 10^{-3})^2}\right\}}$$

$$= \frac{1}{0.1[\pi^2 10^8 + (0.64)10^8]} \text{ F}$$

$$= \frac{10^6}{0.1(10.51 \times 10^8)} \text{ } \mu\text{F}$$

$$= 0.009\,515 \text{ } \mu\text{F or } 9.515 \text{ nF}.$$

(b) Dynamic resistance, $R_D = \dfrac{L}{CR} = \dfrac{100 \times 10^{-3}}{(9.515 \times 10^{-9})(800)} = 13.14 \text{ k}\Omega$

(c) Supply current at resonance, $I_r = \dfrac{V}{R_D} = \dfrac{12}{13.14 \times 10^3} = 0.913 \text{ mA}.$

(d) Q-factor at resonance $= \dfrac{2\pi f_r L}{R} = \dfrac{2\pi(5000)(100 \times 10^{-3})}{800} = 3.93$

Alternatively, Q-factor at resonance $= \dfrac{I_c}{I_r} = \dfrac{V/X_c}{I_r} = \dfrac{2\pi f_r CV}{I_r}$

$$= \frac{2\pi(5000)(9.515 \times 10^{-9})(12)}{0.913 \times 10^{-3}} = 3.93$$

Problem 11 A single-phase motor takes 50 A at a power factor of 0.6 lagging from a 240 V, 50 Hz supply. Determine (a) the current taken by a capacitor connected in parallel with the motor to correct the power factor to unity, and (b) the value of the supply current after power factor correction.

The circuit diagram is shown in *Fig 10(a)*.

(a) A power factor of 0.6 lagging means that $\cos \phi = 0.6$

    i.e. $\phi = \arccos 0.6 = 53° 8'$.

    Hence I_M lags V by $53° 8'$ as shown in *Fig 10(b)*.

    If the power factor is to be improved to unity then the phase difference

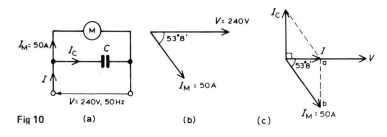

Fig 10   (a)   (b)   (c)

between supply current I and voltage V is $0°$, i.e. I is in phase with V as shown in *Fig 10(c)*. For this to be so, I_C must equal the length ab, such that the phasor sum of I_M and I_C is I.

ab $= I_M \sin 53° 8' = 50(0.8) = 40$ A.

Hence the capacitor current I_c must be 40 A for the power factor to be unity.

(b) Supply current $I = I_M \cos 53° 8' = 50(0.6) = 30$ **A.**

Problem 12 A 400 V alternator is supplying a load of 42 kW at a power factor of 0.7 lagging. Calculate (a) the kVA loading and (b) the current taken from the alternator. (c) If the power factor is now raised to unity find the new kVA loading

(a) Power $= VI \cos \phi = (VI)$ (power factor)

    Hence $VI = \dfrac{\text{power}}{\text{p.f.}} = \dfrac{42 \times 10^3}{0.7} = 60$ kV A

(b) $VI = 60\,000$ V A. Hence $I = \dfrac{60\,000}{V} = \dfrac{60\,000}{400} = 150$ **A**

(c) The kV A loading remains at **60 kV A** irrespective of changes in power factor.

Problem 13 A motor has an output of 4.8 kW, an efficiency of 80% and a power factor of 0.625 lagging when operated from a 240 V, 50 Hz supply. It is required to improve the power factor to 0.95 lagging by connecting a capacitor in parallel with the motor. Determine (a) the current taken by the motor; (b) the supply current after power factor correction; (c) the current taken by the capacitor; (d) the capacitance of the capacitor, and (e) the kV Ar rating of the capacitor.

(a) Efficiency $= \dfrac{\text{power output}}{\text{power input}}$. Hence $\dfrac{80}{100} = \dfrac{4800}{\text{power input}}$

61

Power input $= \dfrac{4800}{0.8} = 6000$ W.

Hence, $6000 = VI_M \cos \phi = (240)(I_M)(0.625)$, since $\cos \phi = $ p.f. $= 0.625$

Thus current taken by the motor, $I_M = \dfrac{6000}{(240)(0.625)} = 40$ A

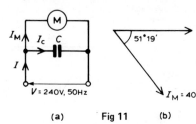

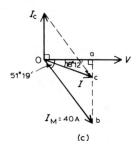

(a)   **Fig 11**   (b)

(c)

The circuit diagram is shown in *Fig 11(a)*.
The phase angle between I_M and V is given by:
$\phi = \arccos 0.625 = 51° 19'$ hence the phasor diagram is as shown in *Fig 11(b)*.

(b) When a capacitor C is connected in parallel with the motor a current I_C flows
which leads V by $90°$. The phasor sum of I_M and I_C gives the supply current I,
and has to be such as to change the circuit power factor to 0.95 lagging, i.e.
a phase angle of arccos 0.95 or $18° 12'$ lagging, as shown in *Fig 11(c)*.
The horizontal component of I_M (shown as oa) $= I_M \cos 51° 19'$
$\qquad\qquad\qquad = 40 \cos 51° 19' = 25$ A
The horizontal component of I (also given by oa) $= I \cos 18° 12' = 0.95\,I$
Equating the horizontal components gives: $25 = 0.95\,I$

Hence the supply current after p.f. correction, $I = \dfrac{25}{0.95} = 26.32$ A

(c) The vertical component of I_M (shown as ab) $= I_M \sin 51° 19'$
$\qquad\qquad\qquad = 40 \sin 51° 19' = 31.22$ A
The vertical component of I (shown as ac) $= I \sin 18° 12'$
$\qquad\qquad\qquad = 26.32 \sin 18° 12' = 8.22$ A
The magnitude of the capacitor current I_C (shown as bc) is given by
ab−ac, i.e. $31.22 - 8.22 = \mathbf{23}$ A

(d) Current $I_C = \dfrac{V}{X_C} = \dfrac{V}{\dfrac{1}{2\pi f C}} = 2\pi f C V,$

from which $\quad C = \dfrac{I_C}{2\pi f V} = \dfrac{23}{2\pi (50)(240)}$ F $= \mathbf{305\ \mu F}$

(e) kV Ar rating of the capacitor $= \dfrac{VI_C}{1000} = \dfrac{(240)(23)}{1000} = \mathbf{5.52\ kV\ Ar}$

In this problem the supply current has been reduced from 40 A to 26.32 A
without altering the current or power taken by the motor. This means that the
size of generating plant and the cross-sectional area of conductors supplying both
the factory and the motor can be less—with an obvious saving in cost.

Problem 14 A 250 V, 50 Hz single-phase supply feeds the following loads (i) incandescent lamps taking a current of 10 A at unity power factor; (ii) fluorescent lamps taking 8 A at a power factor of 0.7 lagging; (iii) a 3 kVA motor operating at full load and at a power factor of 0.8 lagging and (iv) a static capacitor. Determine, for the lamps and motor, (a) the total current; (b) the overall power factor and (c) the total power. (d) Find the value of the static capacitor to improve the overall power factor to 0.975 lagging.

A phasor diagram is constructed as shown in *Fig 12(a)*, where 8 A is lagging voltage V by arccos 0.7, i.e. 45.57°, and the motor current is 3000/250, i.e. 12 A lagging V by arccos 0.8, i.e. 36.87°.

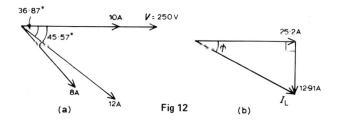

Fig 12

(a) The horizontal component of the currents = $10 \cos 0° + 12 \cos 36.87°$
$+ \ 8 \cos 45.57°$
$= 10+9.6+5.6 = 25.2 \ \text{A}.$
The vertical component of the currents $= 10 \sin 0° + 12 \sin 36.87°$
$+ \ 8 \sin 45.57°$
$= 0+7.2+5.713 = 12.91 \ \text{A}$
From *Fig 12(b)*, total current, $I_L = \sqrt{[(25.2)^2 + (12.91)^2]} = \textbf{28.31 A}$
at a phase angle of $\phi = \arctan \left(\dfrac{12.91}{25.2}\right)$, i.e. 27.13° lagging.

(b) Power factor = $\cos \phi = \cos 27.13° = \textbf{0.890 lagging}.$
(c) Total power, $P = VI_L \cos \phi = (250)(28.31)(0.890) = \textbf{6.3 kW}$
(d) To improve the power factor, a capacitor is connected in parallel with the loads. The capacitor takes a current I_C such that the supply current falls from 28.31 A to I, lagging V by arccos 0.975, i.e. 12.84°. The phasor diagram is shown in *Fig 13*.

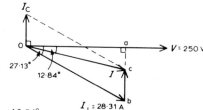

Fig 13

$oa = 28.31 \cos 27.13° = I \cos 12.84°$
Hence $I = \dfrac{28.31 \cos 27.13°}{\cos 12.84°} = 25.84 \ \text{A}$

63

Current $I_C = bc = (ab-ac) = 28.31 \sin 27.13° - 25.84 \sin 12.84°$
$$= 12.91 - 5.742 = 7.168 \text{ A}$$

$$I_C = \frac{V}{X_L} = \frac{V}{\frac{1}{2\pi f C}} = 2\pi f C V$$

Hence capacitance $C = \dfrac{I_C}{2\pi f V} = \dfrac{7.168}{2\pi(50)(250)}$ F $= 91.27 \ \mu$F

Thus to improve the power factor from 0.890 to 0.975 lagging a 91.27 μF capacitor is connected in parallel with the loads.

C. FURTHER PROBLEMS ON PARALLEL A.C. CIRCUITS

(a) SHORT ANSWER PROBLEMS

1 Draw the phasor diagram for a two-branch parallel circuit containing capacitance C in one branch and resistance R in the other, connected across a supply voltage V.

2 Draw the phasor diagram for a two-branch parallel circuit containing inductance L and resistance R in series in one branch, and capacitance C in the other, connected across a supply voltage V.

3 Draw the phasor diagram for a two-branch parallel circuit containing inductance L in one branch and capacitance C in the other for the condition in which inductive reactance is greater than capacitive reactance.

4 State two methods of determining the phasor sum of two currents.

5 State two formulae which may be used to calculate power in a parallel a.c. circuit.

6 State the condition for resonance for a two-branch circuit containing capacitance C in parallel with a coil of inductance L and resistance R.

7 Develop a formula for parallel resonance in terms of resistance R, inductance L and capacitance C.

8 What does the Q-factor of a parallel circuit mean?

9 Develop a formula for the current at resonance in terms of resistance R, inductance L, capacitance C and supply voltage V.

10 What is dynamic resistance? State a formula for dynamic resistance.

11 Explain a simple method of improving the power factor of an inductive circuit.

12 Why is it advantageous to improve power factor?

(b) MULTI-CHOICE PROBLEMS (answers on page 125)

A 2-branch parallel circuit, containing a 10 Ω resistance in one branch and a 100 μF capacitor in the other, has a 120 V, $2/3\pi$ kHz supply connected across it. Determine the quantities stated in *Problems 1 to 8*, selecting the correct answer from the following list.

(a) 24 A; (b) 6 Ω; (c) 7.5 kΩ; (d) 12 A; (e) arctan 3/4 leading; (f) 0.8 leading;
(g) 7.5 Ω; (h) arctan 4/3 leading; (i) 16 A; (j) arctan 5/3 lagging; (k) 1.44 kW;
(l) 0.6 leading; (m) 22.5 Ω; (n) 2.4 kW; (o) arctan 4/3 lagging; (p) 0.6 lagging;
(q) 0.8 lagging; (r) 1.92 kW; (s) 20 A.

1 The current flowing in the resistance.
2 The capacitive reactance of the capacitor.
3 The current flowing in the capacitor.
4 The supply current.
5 The supply phase angle.
6 The circuit impedance.
7 The power consumed by the circuit.
8 The power factor of the circuit.
9 A 2 branch parallel circuit consists of a 15 mH inductance in one branch and a 50 μF capacitor in the other across a 120 V, $1/\pi$ kHz supply. The supply current is:
   (a) 8 A leading by $\pi/2$ rads   (b) 16 A lagging by 90°
   (c) 8 A lagging by 90°   (d) 16 A leading by $\pi/2$ rads.
10 The following statements, taken correct to 2 significant figures, refer to the circuit shown in *Fig 14.* Which are false?
   (a) The impedance of the R–L branch is 5 Ω.
   (b) $I_{LR} = 50$ A; (c) $I_C = 20$ A; (d) $L = 0.80$ H;
   (e) $C = 16$ μF; (f) The 'in-phase' component of the supply current is 30 A
   (g) The 'quadrature' component of the supply current is 40 A (h) $I = 36$ A. (i) Circuit phase angle = 33° 41' leading.
   (j) Circuit impedance = 6.9 Ω;
   (k) Circuit power factor = 0.83 lagging.
   (l) Power consumed = 9.0 kW

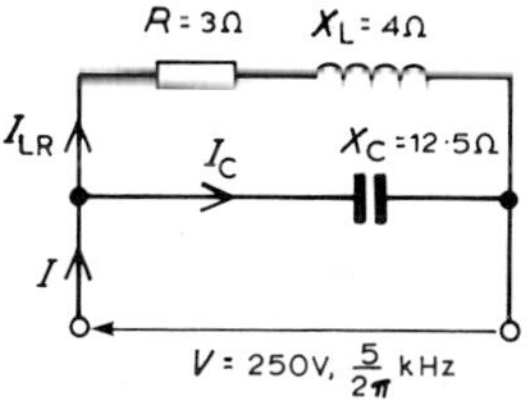

$R = 3\Omega$ $X_L = 4\Omega$

$X_C = 12.5\Omega$

$V = 250$V, $\frac{5}{2\pi}$ kHz

Fig 14

11 Which of the following statements is false?
   (a) The supply current is a minimum at resonance in a parallel circuit.
   (b) The Q-factor at resonance in a parallel circuit is the voltage magnification.
   (c) Improving power factor reduces the current flowing through a system.
   (d) The circuit impedance is a maximum at resonance in a parallel circuit.
12 An LR–C parallel circuit (similar to *Fig 12*) has the following component values: $R = 10$ Ω, $L = 10$ mH, $C = 10$ μF, $V = 100$ V. Which of the following statements is false?
   (a) The resonant frequency f_r is $1.5/\pi$ kHz.
   (b) The current at resonance is 1 A.
   (c) The dynamic resistance is 100 Ω.
   (d) The circuit Q-factor at resonance is 30.

(c) CONVENTIONAL PROBLEMS

1 A 30 Ω resistor is connected in parallel with a pure inductance of 3 mH across a 110 V, 2 kHz supply. Calculate (a) the current in each branch; (b) the circuit current; (c) the circuit phase angle; (d) the circuit impedance; (e) the power consumed, and (f) the circuit power factor.

$$\left[\begin{array}{l} \text{(a) } I_R = 3.67 \text{ A}, I_L = 2.92 \text{ A} \text{ (b) } 4.69 \text{ A} \\ \text{(c) } 38° \text{ } 30' \text{ lagging (d) } 23.45 \text{ Ω (e) } 404 \text{ W} \\ \text{(f) } 0.783 \text{ lagging} \end{array}\right]$$

2 A 40 Ω resistance is connected in parallel with a coil of inductance L and negligible resistance across a 200 V, 50 Hz supply and the supply current is

found to be 8 A. Draw a phasor diagram to scale and determine the inductance of the coil. [102 mH]

3 A 1500 nF capacitor is connected in parallel with a 16 Ω resistor across a 10 V, 10 kHz supply. Calculate (a) the current in each branch; (b) the supply current; (c) the circuit phase angle; (d) the circuit impedance; (e) the power consumed; (f) the apparent power; and (g) the circuit power factor. Draw the phasor diagram.

$$\left[\begin{array}{l} \text{(a) } I_R = 0.625 \text{ A}, I_C = 0.934 \text{ A}; \text{ (b) } 1.13 \text{ A}; \text{ (c) } 56° \ 28' \text{ leading;} \\ \text{(d) } 8.85 \ \Omega; \text{ (e) } 6.25 \text{ W}; \text{ (f) } 11.3 \text{ V A}; \text{ (g) } 0.55 \text{ leading.} \end{array}\right]$$

4 A capacitor C is connected in parallel with a resistance R across a 60 V, 100 Hz supply. The supply current is 0.6 A at a power factor of 0.8 leading. Calculate the value of R and C. [$R = 125 \ \Omega; C = 9.55 \ \mu\text{F}$]

5 An inductance of 80 mH is connected in parallel with a capacitance of 10 μF across a 60 V, 100 Hz supply. Determine (a) the branch currents; (b) the supply current; (c) the circuit phase angle; (d) the circuit impedance and (e) the power consumed.

$$\left[\begin{array}{l} \text{(a) } I_C = 0.377 \text{ A}, I_L = 1.194 \text{ A}; \text{ (b) } 0.817 \text{ A}; \text{ (c) } 90° \text{ lagging;} \\ \text{(d) } 73.44 \ \Omega; \text{ (e) } 0 \text{ W.} \end{array}\right]$$

6 Repeat problem 5 for a supply frequency of 200 Hz.

$$\left[\begin{array}{l} \text{(a) } I_C = 0.754 \text{ A}, I_L = 0.597 \text{ A}; \text{ (b) } 0.157 \text{ A}; \text{ (c) } 90° \text{ leading;} \\ \text{(d) } 382.2 \ \Omega; \text{ (e) } 0 \text{ W.} \end{array}\right]$$

7 A coil of resistance 60 Ω and inductance 318.4 mH is connected in parallel with a 15 μF capacitor across a 200 V, 50 Hz supply. Calculate (a) the current in the coil; (b) the current in the capacitor; (c) the supply current and its phase angle; (d) the circuit impedance; (e) the power consumed; (f) the apparent power and (g) the reactive power. Draw the phasor diagram.

$$\left[\begin{array}{l} \text{(a) } 1.715 \text{ A}; \text{ (b) } 0.943 \text{ A}; \text{ (c) } 1.028 \text{ A at } 30° \ 53' \text{ lagging;} \\ \text{(d) } 194.6 \ \Omega; \text{ (e) } 176.4 \text{ W}; \text{ (f) } 205.6 \text{ VA}; \text{ (g) } 105.5 \text{ VAr.} \end{array}\right]$$

8 A 25 nF capacitor is connected in parallel with a coil of resistance 2 kΩ and inductance 0.20 H across a 100 V, 4 kHz supply. Determine (a) the current in the coil; (b) the current in the capacitor; (c) the supply current and its phase angle (by drawing a phasor diagram to scale, and also by calculation); (d) the circuit impedance; and (e) the power consumed.

$$\left[\begin{array}{l} \text{(a) } 18.48 \text{ mA}; \text{ (b) } 62.83 \text{ mA}; \text{ (c) } 46.17 \text{ mA at } 81° \ 29' \text{ leading;} \\ \text{(d) } 2.166 \text{ k } \Omega; \text{ (e) } 0.684 \text{ W.} \end{array}\right]$$

9 A 0.15 μF capacitor and a pure inductance of 0.01 H are connected in parallel across a 10 V, variable frequency supply. Determine (a) the resonant frequency of the circuit, and (b) the current circulating in the capacitor and inductance. [(a) 4.11 kHz; (b) 38.74 mA]

10 A 30 μF capacitor is connected in parallel with a coil of inductance 50 mH and unknown resistance R across a 120 V, 50 Hz supply. If the circuit has an overall power factor of 1 find (a) the value of R; (b) the current in the coil and (c) the supply current. [(a) 37.7 Ω; (b) 2.94 A; (c) 2.714 A.]

11 A coil of resistance 25 Ω and inductance 150 mH is connected in parallel with a 10 μF capacitor across a 60 V, variable frequency supply. Calculate (a) the resonant frequency; (b) the dynamic resistance; (c) the current at resonance and (d) the Q-factor at resonance. [(a) 127.2 Hz (b) 600 Ω (c) 0.10 A (d) 4.80]

12 A coil having resistance R and inductance 80 mH is connected in parallel with a 5 nF capacitor across a 25 V, 3 kHz supply. Determine for the condition when the current is a minimum, (a) the resistance R of the coil; (b) the dynamic resistance; (c) the supply current; and (d) the Q-factor. [(a) 3.705 kΩ; (b) 4.318 kΩ; (c) 5.79 mA; (d) 0.40.]

13 A 415 V alternator is supplying a load of 55 kW at a power factor of 0.65 lagging. Calculate (a) the kVA loading and (b) the current taken from the alternator. (c) If the power factor is now raised to unity find the new kVA loading.
[(a) 84.6 kVA; (b) 203.9 A; (c) 84.6 kVA.]

14 A coil of resistance 1.5 kΩ and 0.25 H inductance is connected in parallel with a variable capacitance across a 10 V, 8 kHz supply. Calculate (a) the capacitance of the capacitor when the supply current is a minimum; (b) the dynamic resistance, and (c) the supply current. [(a) 1561 pF; (b) 106.8 kΩ; (c) 93.66 μA.]

15 A single phase motor takes 30 A at a power factor of 0.65 lagging from a 240 V, 50 Hz supply. Determine (a) the current taken by the capacitor connected in parallel to correct the power factor to unity; and (b) the value of the supply current after power factor correction. [(a) 22.80 A; (b) 19.5 A.]

16 A motor has an output of 6 kW, an efficiency of 75% and a power factor of 0.64 lagging when operated from a 250 V, 60 Hz supply. It is required to raise the power factor to 0.925 lagging by connecting a capacitor in parallel with the motor. Determine (a) the current taken by the motor; (b) the supply current after power factor correction; (c) the current taken by the capacitor; (d) the capacitance of the capacitor and (e) the kV Ar rating of the capacitor.
[(a) 50 A; (b) 34.59 A; (c) 25.28 A; (d) 268.2 μF; (e) 6.32 kV Ar.]

17 A supply of 250 V, 80 Hz is connected across an inductive load and the power consumed is 2 kW, when the supply current is 10 A. Determine the resistance and inductance of the circuit. What value of capacitance connected in parallel with the load is needed to improve the overall power factor to unity?
[$R = 20\ \Omega$, $L = 29.84$ mH; $C = 47.75\ \mu$F]

18 A 200 V, 50 Hz single-phase supply feeds the following loads: (i) fluorescent lamps taking a current of 8 A at a power factor of 0.9 leading; (ii) incandescent lamps taking a current of 6 A at unity power factor; (iii) a motor taking a current of 12 A at a power factor of 0.65 lagging. Determine the total current taken from the supply and the overall power factor. Find also the value of a static capacitor connected in parallel with the loads to improve the overall power factor to 0.98 lagging. [21.74 A; 0.966 lagging; 21.68 μF]

6 Three-phase systems

A. MAIN POINTS CONCERNED WITH THREE-PHASE SYSTEMS

1   Generation, transmission and distribution of electricity via the National Grid system is accomplished by three-phase alternating currents.

2   The voltage induced by a single coil when rotated in a uniform magnetic field is shown in *Fig 1* and is known as a **single-phase voltage.** Most consumers are

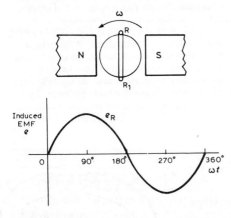

Fig 1

fed by means of a single-phase a.c. supply. Two wires are used, one called the live conductor (usually coloured red) and the other is called the neutral conductor (usually coloured black). The neutral is usually connected via protective gear to earth, the earth wire being coloured green. The standard voltage for a single-phase a.c. supply is 240 V. The majority of single-phase supplies are obtained by connection to a three-phase supply (see *Fig 5*).

3   A **three-phase supply** is generated when three coils are placed 120° apart and the whole rotated in a uniform magnetic field as shown in *Fig 2(a)*. The result is three independent supplies of equal voltages which are each displaced by 120° from each other as shown in *Fig 2(b)*.

4   (i)   The convention adopted to identify each of the phase voltages is: R-red, Y-yellow, and B-blue, as shown in *Fig 2*.

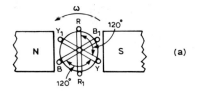

(a)

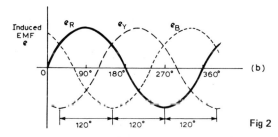

(b)

Fig 2

(ii) The **phase-sequence** is given by the sequence in which the conductors pass the point initially taken by the red conductor. The national standard phase sequence is R, Y, B.

5 A three-phase a.c. supply is carried by three conductors, called 'lines' which are coloured red, yellow and blue. The currents in these conductors are known as line currents (I_L) and the p.d.'s between them are known as line voltages (V_L). A fourth conductor, called the **neutral** (coloured black, and connected through protective devices to earth) is often used with a three-phase supply.

6 If the three-phase windings shown in *Fig 2* are kept independent then six wires are needed to connect a supply source (such as a generator) to a load (such as motor). To reduce the number of wires it is usual to interconnect the three phases. There are two ways in which this can be done, these being: (a) a star connection, and (b) a delta, or mesh, connection. Sources of three-phase supplies, i.e. alternators, are usually connected in star, whereas three-phase transformer windings, motors and other loads may be connected either in star or delta.

7 (i) A **star-connected load** is shown in *Fig 3* where the three line conductors are each connected to a load and the outlets from the loads are joined together at N to form what is termed the **neutral point** or the **star point**.

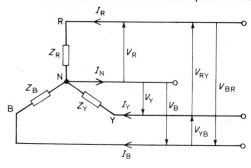

Fig 3

(ii) The voltages, V_R, V_Y and V_B are called **phase voltages** or line to neutral voltages. Phase voltages are generally denoted by V_p.

(iii) The voltages, V_{RY}, V_{YB} and V_{BR} are called **line voltages**.

(iv) From *Fig 3* it can be seen that the phase currents (generally denoted by I_p) are equal to their respective line currents I_R, I_Y and I_B, i.e. for a star connection:

$$I_L = I_p$$

(v) For a balanced system: $\quad I_R = I_Y = I_B, \quad V_R = V_Y = V_B$
$$V_{RY} = V_{YB} = V_{BR}, \quad Z_R = Z_Y = Z_B$$
and the current in the neutral conductor, $\quad I_N = 0$.

When a star connected system is balanced, then the neutral conductor is unnecessary and is often omitted.

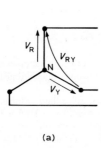

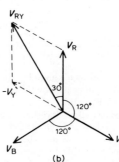

    (a)   **Fig 4**   (b)

(vi) The line voltage, V_{RY}, shown in *Fig 4(a)* is given by $V_{RY} = V_R - V_Y$. (V_Y is negative since it is in the opposite direction to V_{RY}.) In the phasor diagram of *Fig 4(b)*, phasor V_Y is reversed (shown by the broken line) and then added phasorially to V_R (i.e. $V_{RY} = V_R + (-V_Y)$). By trigonometry, or by measurement, $V_{RY} = \sqrt{3}V_R$, i.e. for a balanced star connection:

$$V_L = \sqrt{3} \ V_p$$

(See *Problem 6* for a complete phasor diagram of a star-connected system.)

(vii) The star connection of the three phases of a supply, together with a neutral conductor, allows the use of two voltages—the phase voltage and the line voltage. A 4-wire system is also used when the load is not balanced. The standard electricity supply to consumers in Great Britain is 415/240 V, 50 Hz, 3-phase, 4-wire alternating current, and a diagram of connections is shown in *Fig 5*.

8 (i) **A delta (or mesh) connected load** is shown in *Fig 6* where the end of one load is connected to the start of the next load.

(ii) From *Fig 6*, it can be seen that the line voltages V_{RY}, V_{YB} and V_{BR} are the respective phase voltages, i.e. for a delta connection:

$$V_L = V_p$$

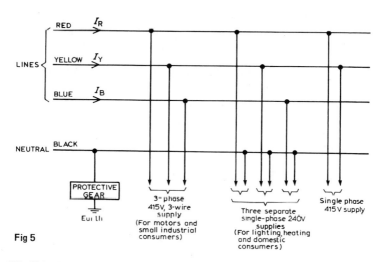

Fig 5

(iii) Using Kirchhoff's current law in *Fig 6*, $I_R = I_{RY} - I_{BR} = I_{RY} + (-I_{BR})$. From the phasor diagram shown in *Fig 7*, by trigonometry or by measurement, $I_R = \sqrt{3}\,I_{RY}$, i.e. for a delta connection:

$$\boxed{I_L = \sqrt{3}\,I_p}$$

9 The power dissipated in a three-phase load is given by the sum of the power dissipated in each phase. If a load is balanced then the total power P is given by:

$P = 3 \times$ power consumed by one phase.

The power consumed in one phase $= I_p^2 R_p$ or $V_p I_p \cos\phi$ (where ϕ is the phase angle between V_p and I_p).

For a star connection $V_p = \dfrac{V_L}{\sqrt{3}}$ and $I_p = I_L$ hence $P = 3\left(\dfrac{V_L}{\sqrt{3}}\right) I_L \cos\phi$

$$= \sqrt{3}\,V_L I_L \cos\phi.$$

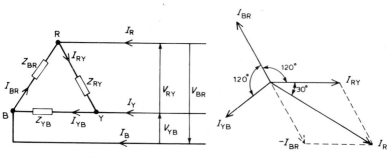

Fig 6   Fig 7

71

For a delta connection, $V_p = V_L$ and $I_p = \dfrac{I_L}{\sqrt{3}}$ hence $P = 3V_L\left(\dfrac{I_L}{\sqrt{3}}\right)\cos\phi$

$$= \sqrt{3}\, V_L I_L \cos\phi.$$

Hence for either a star or a delta balanced connection the total power P is given by:

$P = \sqrt{3}\, V_L I_L \cos\phi$ **watts** or $P = 3 I_p^2 R_p$ **watts.**

Total volt-amperes, $S = \sqrt{3}\, V_L I_L$ **volt-amperes.**

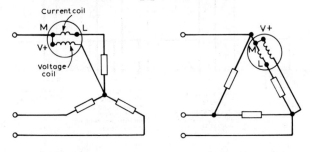

Fig 8

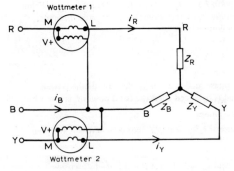

Fig 9

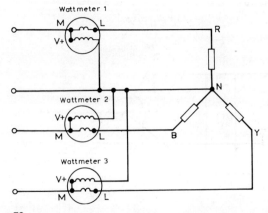

Fig 10

10 Power in three-phase loads may be measured by the following methods:
  (i) **One-wattmeter method for a balanced load.**
      Wattmeter connections for both star and delta are shown in *Fig 8*.
      Total power = 3 × wattmeter reading
  (ii) **Two-wattmeter method for balanced or unbalanced loads.**
      A connection diagram for this method is shown in *Fig 9* for a star-connected load. Similar connections are made for a delta-connected load.
      Total power = sum of wattmeter readings = $P_1 + P_2$.
      The power factor may be determined from:

      $$\tan \phi = \sqrt{3} \frac{(P_1 - P_2)}{(P_1 + P_2)} \quad \text{(see Problems 11 and 14 to 17).}$$

      It is possible, depending on the load power factor, for one wattmeter to have to be 'reversed' to obtain a reading. In this case it is taken as a negative reading (see *Problem 16*).
  (iii) **Three-wattmeter method for a three-phase, 4-wire system for balanced and unbalanced loads** (see *Fig 10*)
      Total power = $P_1 + P_2 + P_3$
11 (i) Loads connected in delta dissipate three times more power than when connected in star to the same supply.
   (ii) For the same power, the phase currents must be the same for both delta and star connections (since power = $3 I_p^2 R_p$), hence the line current in the delta-connected system is greater than the line current in the corresponding star-connected system. To achieve the same phase current in a star-connected system as in a delta-connected system, the line voltage in the star system is $\sqrt{3}$ times the line voltage in the delta system.
      Thus for a given power transfer, a delta system is associated with larger line currents (and thus larger conductor cross-sectional area) and a star system is associated with a larger line voltage (and thus greater insulation).
12 **Advantages of three-phase systems** over single-phase supplies include:
   (i) For a given amount of power transmitted through a system, the three-phase system requires conductors with a smaller cross-sectional area. This means a saving of copper (or aluminium) and thus the original installation costs are less.
   (ii) Two voltages are available (see para. 7).
   (iii) Three-phase motors are very robust, relatively cheap, generally smaller, have self-starting properties, provide a steadier output and require little maintenance compared with single-phase motors.

B. WORKED PROBLEMS ON THREE-PHASE SYSTEMS

Problem 1 Three loads, each of resistance 30 Ω, are connected in star to a 415 V, 3-phase supply. Determine (a) the system phase voltage; (b) the phase current and (c) the line current.

A '415 V, 3-phase supply' means that 415 V is the line voltage, V_L.
(a) For a star connection, $V_L = \sqrt{3} \, V_p$

Hence phase voltage, $V_p = \dfrac{V_L}{\sqrt{3}} = \dfrac{415}{\sqrt{3}} = \textbf{239.6 V or 240 V}$ correct to 3 significant figures.

(b) Phase current, $I_p = \dfrac{V_p}{R_p} = \dfrac{240}{30} = 8$ A

(c) For a star connection, $I_p = I_L$
   Hence the line current, $I_L = 8$ A

Problem 2 A star-connected load consists of three identical coils each of resistance 30 Ω and inductance 127.3 mH. If the line current is 5.08 A, calculate the line voltage if the supply frequency is 50 Hz.

Inductive reactance $X_L = 2\pi fL = 2\pi(50)(127.3 \times 10^{-3}) = 40$ Ω
Impedance of each phase $Z_p = \sqrt{(R^2 + X_L^2)} = \sqrt{(30^2 + 40^2)} = 50$ Ω

For a star connection $I_L = I_p = \dfrac{V_p}{Z_p}$

Hence phase voltage $V_p = I_p Z_p = (5.08)(50) = 254$ V
Line voltage $V_L = \sqrt{3} V_p = \sqrt{3}\,(254) = \mathbf{440}$ **V**

Problem 3 The three coils in *Problem 2* are now connected in delta to the 440 V, 50 Hz, 3-phase supply. Determine (a) the phase current and (b) the line current.

Phase impedance, $Z_p = 50$ Ω (as above) and for a delta connection $V_p = V_L$

(a) Phase current, $I_p = \dfrac{V_p}{Z_p} = \dfrac{V_L}{Z_p} = \dfrac{440}{50} = \mathbf{8.8}$ **A**

(b) For a delta connection, $I_L = \sqrt{3}\,I_p = \sqrt{3}\,(8.8) = \mathbf{15.24}$ **A**

Thus when the load is connected in delta, three times the line current is taken from the supply than is taken if connected in star.

Problem 4 Three identical capacitors are connected in delta to a 415 V, 50 Hz, 3-phase supply. If the line current is 15 A, determine the capacitance of each of the capacitors.

For a delta connection $I_L = \sqrt{3}\,I_p$

Hence phase current $I_p = \dfrac{I_L}{\sqrt{3}} = \dfrac{15}{\sqrt{3}} = 8.66$ A

Capacitive reactance per phase, $X_C = \dfrac{V_p}{I_p} = \dfrac{V_L}{I_p}$ (since for a delta connection $V_L = V_p$).

Hence $X_C = \dfrac{415}{8.66} = 47.92$ Ω

$X_C = \dfrac{1}{2\pi fC}$, from which capacitance, $C = \dfrac{1}{2\pi f X_C} = \dfrac{1}{2\pi(50)(47.92)}$ F

$= \mathbf{66.43}\ \boldsymbol{\mu}$**F**

Problem 5 Three coils each having resistance 3 Ω and inductive reactance 4 Ω are connected (i) in star and (ii) in delta to a 415 V, 3-phase supply. Calculate for each connection (a) the line and phase voltages and (b) the phase and line currents.

(i) **For a star connection:** $I_L = I_p$ and $V_L = \sqrt{3} \, V_p$

    (a) A 415 V, 3-phase supply means that the line voltage, $V_L = 415$ V

         Phase voltage, $V_p = \dfrac{V_L}{\sqrt{3}} = \dfrac{415}{\sqrt{3}} = 240$ V

    (b) Impedance per phase, $Z_p = \sqrt{(R^2 + X_L^2)} = \sqrt{(3^2 + 4^2)} = 5 \ \Omega$

         Phase current, $I_p = \dfrac{V_p}{Z_p} = \dfrac{240}{5} = 48$ A

         Line current, $I_L = I_p = \mathbf{48}$ **A**

(ii) **For a delta connection:** $V_L = V_p$ and $I_L = \sqrt{3} \, I_p$

    (a) Line voltage, $V_L = 415$ V

         Phase voltage, $V_p = V_L = 415$ V

    (b) Phase current, $I_p = \dfrac{V_p}{Z_p} = \dfrac{415}{5} = 83$ A

         Line current, $I_L = \sqrt{3} \, I_p = \sqrt{3}(83) = \mathbf{144}$ **A**

Problem 6 A balanced, three-wire, star-connected, 3-phase load has a phase voltage of 240 V, a line current of 5 A and a lagging power factor of 0.966. Draw the complete phasor diagram.

The phasor diagram is shown in *Fig 11*.
Procedure to construct the phasor diagram:

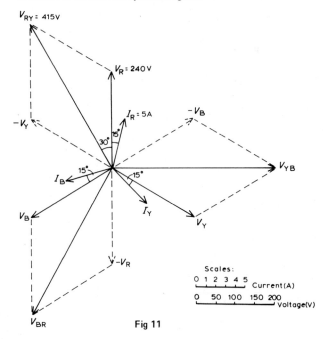

Fig 11

(i) Draw $V_R = V_Y = V_B = 240$ V and spaced $120°$ apart. (Note that V_R is shown vertically upwards—this however is immaterial for it may be drawn in any direction.)

(ii) Power factor $= \cos \phi = 0.966$ lagging. Hence the load phase angle is given by arccos 0.966, i.e. $15°$ lagging. Hence $I_R = I_Y = I_B = 5$ A, lagging V_R, V_Y and V_B respectively by $15°$.

(iii) $V_{RY} = V_R - V_Y$ (phasorially). Hence V_Y is reversed and added phasorially to V_R. By measurement, $V_{RY} = 415$ V (i.e. $\sqrt{3}(240)$) and leads V_R by $30°$. Similarly, $V_{YB} = V_Y - V_B$ and $V_{BR} = V_B - V_R$.

Problem 7 Three 12 Ω resistors are connected in star to a 415 V, 3-phase supply. Determine the total power dissipated by the resistors.

Power dissipated, $\qquad P = \sqrt{3} \, V_L I_L \cos \phi$ or $P = 3 I_p^2 R_p$.

Line voltage, $V_L = 415$ V and phase voltage $V_p = \dfrac{415}{\sqrt{3}} = 240$ V

$\qquad\qquad\qquad\qquad$ (since the resistors are star-connected)

Phase current, $I_p = \dfrac{V_p}{Z_p} = \dfrac{V_p}{R_p} = \dfrac{240}{12} = 20$ A

For a star connection, $I_L = I_p = 20$ A

For a purely resistive load, the power factor $= \cos \phi = 1$

Hence power $P = \sqrt{3} \, V_L I_L \cos \phi = \sqrt{3} \,(415)(20)(1) = $ **14.4 kW**

$\quad$ or power $P = 3 I_p^2 R_p = 3(20)^2(12) = $ **14.4 kW**

Problem 8 The input power to a 3-phase a.c. motor is measured as 5 kW. If the voltage and current to the motor are 400 V and 8.6 A respectively, determine the power factor of the system.

Power, $P = 5000$ W; Line voltage $V_L = 400$ V; Line current, $I_L = 8.6$ A
Power, $P = \sqrt{3} \, V_L I_L \cos \phi$

Hence power factor $= \cos \phi = \dfrac{P}{\sqrt{3} \, V_L I_L} = \dfrac{5000}{\sqrt{3}(400)(8.6)} = $ **0.839**

Problem 9 Three identical coils, each of resistance 10 Ω and inductance 42 mH are connected (a) in star and (b) in delta to a 415 V, 50 Hz, 3-phase supply. Determine the total power dissipated in each case.

(a) **Star-connection**

Inductive reactance $\quad X_L = 2\pi f L = 2\pi(50)(42 \times 10^{-3}) = 13.19 \, \Omega$

Phase impedance $\quad Z_p = \sqrt{(R^2 + X_L^2)} = \sqrt{(10^2 + 13.19^2)} = 16.55 \, \Omega$

Line voltage $V_L = 415$ V and phase voltage, $V_p = \dfrac{V_L}{\sqrt{3}} = \dfrac{415}{\sqrt{3}} = 240$ V

Phase current $\quad I_p = \dfrac{V_p}{Z_p} = \dfrac{240}{16.55} = 14.50$ A

Line current $\quad I_L = I_p = 14.50$ A

Power factor $= \cos \phi = \dfrac{R_p}{Z_p} = \dfrac{10}{16.55} = 0.6042$ lagging

Power dissipated $P = \sqrt{3}\,V_L I_L \cos\phi = \sqrt{3}(415)(14.50)(0.6042) = \textbf{6.3 kW}$
(Alternatively $\quad P - 3I_p{}^2 R_p - 3(14.50)^2(10) = \textbf{6.3 kW}$)

(b) **Delta-connection**

$V_L = V_p = 415$ V; $Z_p = 16.55\ \Omega$; $\cos\phi = 0.6042$ lagging (from above).

Phase current $\quad I_p = \dfrac{V_p}{Z_p} = \dfrac{415}{16.55} = 25.08$ A

Line current $\quad I_L = \sqrt{3}I_p = \sqrt{3}(25.08) = 43.44$ A
Power dissipated $P = \sqrt{3}\,V_L I_L \cos\phi = \sqrt{3}(415)(43.44)(0.6042) = \textbf{18.87 kW}$
(Alternatively $\quad P = 3I_p{}^2 R_p = 3(25.08)^2(10) = \textbf{18.87 kW}$)

Hence loads connected in delta dissipate three times the power than when connected in star and also take a line current three times greater.

Problem 10 A 415 V, 3-phase a.c. motor has a power output of 12.75 kW and operates at a power factor of 0.77 lagging and with an efficiency of 85%. If the motor is delta-connected, determine (a) the power input; (b) the line current and (c) the phase current.

(a) Efficiency $= \dfrac{\text{power output}}{\text{power input}}$. Hence $\dfrac{85}{100} = \dfrac{12\,750}{\text{power input}}$

from which, power input $= \dfrac{12\,750 \times 100}{85} = 15\,000$ W or **15 kW**

(b) Power, $P = \sqrt{3}\,V_L I_L \cos\phi$, hence line current, $I_L = \dfrac{P}{\sqrt{3}\,V_L \cos\phi}$

$= \dfrac{15\,000}{3(415)(0.77)} = \textbf{27.10 A}$

(c) For a delta connection, $I_L = \sqrt{3}I_p$. Hence phase current, $I_p = \dfrac{I_L}{\sqrt{3}} = \dfrac{27.10}{\sqrt{3}}$

$= \textbf{15.65 A}$

Problem 11 (a) Show that the total power in a 3-phase, 3-wire system using the two-wattmeter method of measurement is given by the sum of the wattmeter readings. Draw a connection diagram.
(b) Draw a phasor diagram for the two-wattmeter method for a balanced load.
(c) Use the phasor diagram of part (b) to derive a formula from which the power factor of a 3-phase system may be determined using only the watt-meter readings.

(a) A connection diagram for the two-wattmeter method of power measurement is shown in *Fig 12* for a star-connected load.
Total instantaneous power, $p = e_R i_R + e_Y i_Y + e_B i_B$ and in any 3 phase system $i_R + i_Y + i_B = 0$. Hence $i_B = -i_R - i_Y$.
Thus, $\quad p = e_R i_R + e_Y i_Y + e_B(-i_R - i_Y)$
$\quad\quad = (e_R - e_B)i_R + (e_Y - e_B)i_Y$
However, $(e_R - e_B)$ is the p.d. across wattmeter 1 in *Fig 12* and $(e_Y - e_B)$ is the p.d. across wattmeter 2.
Hence total instantaneous power

$p = (\text{wattmeter 1 reading}) + (\text{wattmeter 2 reading}) = p_1 + p_2.$

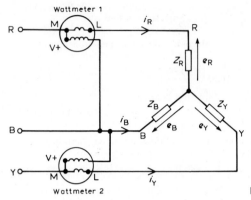

Fig 12

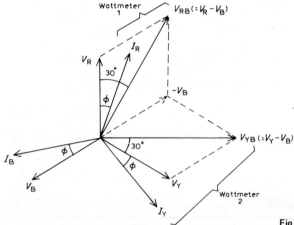

Fig 13

The moving systems of the wattmeters are unable to follow the variations which take place at normal frequencies and they indicate the mean power taken over a cycle. Hence the total power, $P = P_1 + P_2$ for balanced or unbalanced loads.

(b) The phasor diagram for the two-wattmeter method for a balanced load having a lagging current is shown in *Fig 13*, where $V_{RB} = V_R - V_B$ and $V_{YB} = V_Y - V_B$ (phasorially).

(c) Wattmeter 1 reads $V_{RB}I_R \cos(30° - \phi) = P_1$
Wattmeter 2 reads $V_{YB}I_Y \cos(30° + \phi) = P_2$

$$\frac{P_1}{P_2} = \frac{V_{RB}I_R \cos(30° - \phi)}{V_{YB}I_Y \cos(30° + \phi)} = \frac{\cos(30° - \phi)}{\cos(30° + \phi)} \, ,$$

since $I_R = I_Y$ and $V_{RB} = V_{YB}$ for a balanced load.

Hence $\dfrac{P_1}{P_2} = \dfrac{\cos 30° \cos \phi + \sin 30° \sin \phi}{\cos 30° \cos \phi - \sin 30° \sin \rho}$ (from compound angle formulae)

Dividing throughout by $\cos 30° \cos \phi$ gives:

$$\dfrac{P_1}{P_2} = \dfrac{1 + \tan 30° \tan \phi}{1 - \tan 30° \tan \phi} = \dfrac{1 + \dfrac{1}{\sqrt 3} \tan \phi}{1 - \dfrac{1}{\sqrt 3} \tan \phi} \ , \ \left(\text{since } \dfrac{\sin \phi}{\cos \phi} = \tan \phi\right)$$

Cross-multiplying gives: $P_1 - \dfrac{P_1}{\sqrt 3} \tan \phi = P_2 + \dfrac{P_2}{\sqrt 3} \tan \phi.$

Hence $P_1 - P_2 = (P_1 + P_2) \dfrac{\tan \phi}{\sqrt 3}$

from which $\mathbf{\tan \phi = \sqrt 3 \left(\dfrac{P_1 - P_2}{P_1 + P_2}\right).}$

ϕ, $\cos \phi$ and thus power factor can be determined from this formula.

Problem 12 A 400 V, 3-phase star connected alternator supplies a delta-connected load, each phase of which has a resistance of 30 Ω and inductive reactance 40 Ω. Calculate (a) the current supplied by the alternator and (b) the output power and 4 kVA of the alternator, neglecting losses in the line between the alternator and load.

A circuit diagram of the alternator and load is shown in *Fig 14*.

(a) Considering the load: Phase current, $I_p = \dfrac{V_p}{Z_p}$

$V_p = V_L$ for a delta connection. Hence $V_p = 400$ V
Phase impedance, $Z_p = \sqrt{(R_p{}^2 + X_L{}^2)} = \sqrt{(30^2 + 40^2)} = 50$ Ω
Hence $I_p = \dfrac{V_p}{Z_p} = \dfrac{400}{50} = 8$ A

For a delta-connection, line current, $I_L = \sqrt 3 I_p = \sqrt 3 (8) = 13.86$ A
Hence 13.86 A is the current supplied by the alternator.

(b) Alternator output power is equal to the power dissipated by the load,

i.e. $P = \sqrt 3 V_L I_L \cos \phi$, where $\cos \phi = \dfrac{R_p}{Z_p} = \dfrac{30}{50} = 0.6$

Hence $P = \sqrt 3 (400)(13.86)(0.6)$
$\quad\quad = \textbf{5.76 kW}$

Alternator output
kVA, $S = \sqrt 3 V_L I_L$
$\quad = \sqrt 3 (400)(13.86)$
$\quad = \textbf{9.60 kVA}$

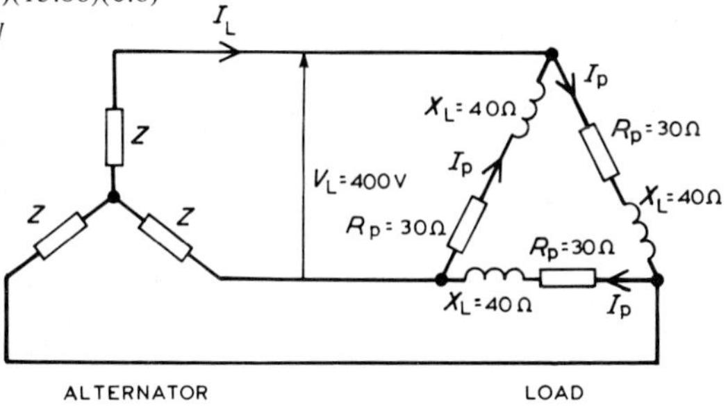

Fig 14   ALTERNATOR   LOAD

79

Problem 13 Each phase of a delta-connected load comprises a resistance of 30 Ω and an 80 µF capacitor in series. The load is connected to a 400 V, 50 Hz, 3-phase supply. Calculate (a) the phase current; (b) the line current; (c) the total power dissipated and (d) the kVA rating of the load. Draw the complete phasor diagram for the load.

(a) Capacitive reactance, $X_C = \dfrac{1}{2\pi f C} = \dfrac{1}{2\pi(50)(80 \times 10^{-6})} = 39.79\ \Omega$

Phase impedance, $Z_p = \sqrt{(R_p{}^2 + X_C{}^2)} = \sqrt{(30^2 + 39.79^2)} = 49.83\ \Omega$

Power factor $= \cos\phi = \dfrac{R_p}{Z_p} = \dfrac{30}{49.83} = 0.602$. Hence $\phi = \arccos 0.602$
$= 52° 59'$ leading.

Phase current, $I_p = \dfrac{V_p}{Z_p}$ and $V_p = V_L$ for a delta connection.

Hence $I_p = \dfrac{400}{49.83} = \mathbf{8.027\ A}$

(b) Line current $I_L = \sqrt{3}\, I_p$ for a delta-connection.
Hence $I_L = \sqrt{3}\,(8.027) = \mathbf{13.90\ A}$

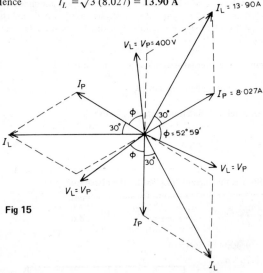

Fig 15

(c) Total power dissipated, $P = \sqrt{3}\, V_L I_L \cos\phi = \sqrt{3}(400)(13.90)(0.602)$
$= \mathbf{5.797\ kW}$
(d) Total kVA, $S = \sqrt{3}\, V_L I_L = \sqrt{3}(400)(13.90) = \mathbf{9.630\ kVA}$
The phasor diagram for the load is shown in *Fig 15*.

Problem 14 Two wattmeters are connected to measure the input power to a balanced 3-phase load by the two-wattmeter method. If the instrument readings are 8 kW and 4 kW, determine (a) the total power input and (b) the load power factor.

With reference to para. 10(ii):

(a) Total input power $P = P_1 + P_2 = 8 + 4 = 12 \text{ kW}$

(b) $\tan\phi = \sqrt{3}\left(\dfrac{P_1 - P_2}{P_1 + P_2}\right) = \sqrt{3}\left(\dfrac{8-4}{8+4}\right) = \sqrt{3}\left(\dfrac{4}{12}\right) = \sqrt{3}\left(\dfrac{1}{3}\right) = \dfrac{1}{\sqrt{3}}$

Hence $\phi = \arctan\dfrac{1}{\sqrt{3}} = 30°$

Power factor $= \cos\phi = \cos 30° = \textbf{0.866}$

Problem 15 Two wattmeters connected to a 3-phase motor indicate the total power input to be 12 kW. The power factor is 0.6. Determine the reading of each wattmeter.

If the two wattmeters indicate P_1 and P_2 respectively then $P_1 + P_2 = 12 \text{ kW}$　(1)

$\tan\phi = \sqrt{3}\left(\dfrac{P_1 - P_2}{P_1 + P_2}\right)$　and power factor $= 0.6 = \cos\phi$

Angle $\phi = \arccos 0.6 = 53°\ 8'$ and $\tan 53°\ 8' = 1.3333$

Hence $1.3333 = \dfrac{\sqrt{3}\,(P_1 - P_2)}{12}$ from which $P_1 - P_2 = \dfrac{12\,(1.3333)}{\sqrt{3}} = 9.237 \text{ kW}$　(2)

Adding equations (1) and (2) gives: $2P_1 = 21.237$, i.e. $P_1 = \dfrac{21.237}{2} = 10.62 \text{ kW}$

Hence **wattmeter 1 reads 10.62 kW**
From equation (1), **wattmeter 2 reads** $(12 - 10.62) = 1.38 \text{ kW}$

Problem 16 Two wattmeters indicate 10 kW and 3 kW respectively when connected to measure the input power to a 3-phase balanced load, the reverse switch being operated on the meter indicating the 3 kW reading. Determine (a) the input power and (b) the load power factor.

Since the reversing switch on the wattmeter had to be operated the 3 kW reading is taken as -3 kW.

(a) Total input power, $P = P_1 + P_2 = 10 + (-3) = \textbf{7 kW}$

(b) $\tan\phi = \sqrt{3}\left(\dfrac{P_1 - P_2}{P_1 + P_2}\right) = \sqrt{3}\left(\dfrac{10 - (-3)}{10 + (-3)}\right) = \sqrt{3}\left(\dfrac{13}{7}\right) = 3.2167$

Angle $\phi = \arctan 3.2167 = 72°\ 44'$.
Power factor $= \cos\phi = \cos 72°\ 44' = \textbf{0.297}$

Problem 17 Three similar coils, each having a resistance of 8 Ω and an inductive reactance of 8 Ω are connected (a) in star and (b) in delta, across a 415 V, 3-phase supply. Calculate for each connection the readings on each of two wattmeters connected to measure the power by the two-wattmeter method.

(a) **Star connection**: $V_L = \sqrt{3}\,V_p$ and $I_L = I_p$

Phase voltage, $V_p = \dfrac{V_L}{\sqrt{3}} = \dfrac{415}{\sqrt{3}}$ and phase impedance, $Z_p = \sqrt{(R_p^2 + X_L^2)}$
$= \sqrt{(8^2 + 8^2)}$
$= 11.31\ \Omega$

Hence phase current, $I_p = \dfrac{V_p}{Z_p} = \dfrac{415/\sqrt{3}}{11.31} = 21.18 \text{ A}$

Total power $P = 3I_p{}^2R_p = 3(21.18)^2(8) = 10\ 766$ W

If wattmeter readings are P_1 and P_2 then $P_1 + P_2 = 10\ 766$ (1)

Since $R_p = 8\ \Omega$ and $X_L = 8\ \Omega$, then phase angle $\phi = 45°$ (from impedance triangle)

$$\tan\phi = \sqrt{3}\left(\frac{P_1 - P_2}{P_1 + P_2}\right), \text{ hence } \tan 45° = \frac{\sqrt{3}(P_1 - P_2)}{10\ 766}$$

from which $P_1 - P_2 = \dfrac{10\ 766\ (1)}{\sqrt{3}} = 6216$ W   (2)

Adding equations (1) and (2) gives: $2P_1 = 10\ 766 + 6216 = 16\ 982$ W

Hence $P_1 = 8491$ W

From equation (1), $P_2 = 10\ 766 - 8491 = 2275$ W

When the coils are star-connected the wattmeter readings are thus 8.491 kW and 2.275 kW.

(b) **Delta connection:** $V_L = V_p$ and $I_L = \sqrt{3}I_p$

Phase current, $I_p = \dfrac{V_p}{Z_p} = \dfrac{415}{11.31} = 36.69$ A

Total power $P = 3I_p{}^2R_p = 3(36.69)^2(8) = 32\ 310$ W

Hence $P_1 + P_2 = 32\ 310$ W   (3)

$$\tan\phi = \sqrt{3}\left(\frac{P_1 - P_2}{P_1 + P_2}\right) \text{ thus } 1 = \frac{\sqrt{3}(P_1 - P_2)}{32\ 310}$$

from which $P_1 - P_2 = \dfrac{32\ 310}{\sqrt{3}} = 18\ 650$ W   (4)

Adding equations (3) and (4) gives: $2P_1 = 50\ 960$, from which $P_1 = 25\ 480$ W

From equation (3), $P_2 = 32\ 310 - 25\ 480 = 6830$ W

When the coils are delta-connected the wattmeter readings are thus 25.48 kW and 6.83 kW.

C. FURTHER PROBLEMS ON THREE-PHASE SYSTEMS

(a) SHORT ANSWER PROBLEMS

1 Explain briefly how a three-phase supply is generated.

2 State the national standard phase sequence for a three-phase supply.

3 State the two ways in which phases of a three-phase supply can be inter-connected to reduce the number of conductors used compared with three, single-phase systems.

4 State the relationships between line and phase currents and line and phase voltages for a star-connected system.

5 When may the neutral conductor of a star-connected system be omitted?

6 State the relationships between line and phase currents and line and phase voltages for a delta-connected system.

7 What is the standard electricity supply to consumers in Great Britain?

8 State two formulae for determining the power dissipated in the load of a 3-phase balanced system.

9 By what methods may power be measured in a three-phase system?

10 State a formula from which power factor may be determined for a balanced system when using the two-wattmeter method of power measurement.

11 Loads connected in star dissipate . . . . the power dissipated when connected in delta and fed from the same supply.

12 Name three advantages of three-phase systems over single-phase systems.

Three loads, each of 10 Ω resistance, are connected in star to a 400 V, 3-phase supply. Determine the quantities stated in Problems 1 to 5, selecting the correct answers from the following list.

(a) $\dfrac{40}{\sqrt{3}}$ A; (b) $\sqrt{3}$ (16) kW; (c) $\dfrac{400}{\sqrt{3}}$ V; (d) $\sqrt{3}$ (40) A; (e) $\sqrt{3}$ (400) V;

(f) 16 kW; (g) 400 V; (k) 48 kW; (i) 40 A.

1 Line voltage.
2 Phase voltage.
3 Phase current.
4 Line current.
5 Total power dissipated in the load.
6 Which of the following statements is false?
   (a) For the same power, loads connected in delta have a higher line voltage and a smaller line current than loads connected in star.
   (b) When using the two-wattmeter method of power measurement the power factor is unity when the wattmeter readings are the same.
   (c) a.c. may be distributed using a single-phase system with 2 wires, a three-phase system with 3 wires or a three-phase system with 4 wires.
   (d) The national standard phase sequence for a three-phase supply is R, Y, B.

Three loads, each of resistance 16 Ω and inductive reactance 12 Ω are connected in delta to a 400 V, 3-phase supply. Determine the quantities stated in problems 7 to 12, selecting the correct answer from the following list.

(a) 4 Ω; (b) $\sqrt{3}$ (400) V; (c) $\sqrt{3}$ (6.4) kW; (d) 20 A; (e) 6.4 kW; (f) $\sqrt{3}$ (20) A;

(g) 20 Ω; (h) $\dfrac{20}{\sqrt{3}}$ A; (i) $\dfrac{400}{\sqrt{3}}$ V; (j) 19.2 kW; (k) 100 A; (l) 400 V; (m) 28 Ω.

7 Phase impedance.
8 Line voltage.
9 Phase voltage.
10 Phase current.
11 Line current.
12 Total power dissipated in the load.

1 Three loads, each of resistance 50 Ω are connected in star to a 400 V, 3-phase supply. Determine (a) the phase voltage; (b) the phase current and (c) the line current.   [(a) 231 V; (b) 4.62 A; (c) 4.62 A.]

2 If the loads in *Problem 1* are connected in delta to the same supply determine (a) the phase voltage; (b) the phase current and (c) the line current.
    [(a) 400 V; (b) 8 A; (c) 13.86 A.]

3 A star-connected load consists of three identical coils, each of inductance

159.2 mH and resistance 50 Ω. If the supply frequency is 50 Hz and the line current is 3 A determine (a) the phase voltage and (b) the line voltage.

[(a) 212 V; (b) 367 V.]

4 Obtain a relationship between the line and phase voltages and line and phase current for a delta connected system. Three inductive loads each of resistance 75 Ω and inductance 318.4 mH are connected in delta to a 415 V, 50 Hz, 3-phase supply. Determine (a) the phase voltage; (b) the phase current, and (c) the line current. [(a) 415 V; (b) 3.32 A; (c) 5.75 A.]

5 Three identical capacitors are connected (a) in star, (b) in delta to a 400 V, 50 Hz 3-phase supply. If the line current is 12 A determine in each case the capacitance of each of the capacitors. [(a) 165.4 μF; (b) 55.13 μF.]

6 Three coils each having resistance 6 Ω and inductance L H are connected (a) in star and (b) in delta, to a 415 V, 50 Hz, 3-phase supply. If the line current is 30 A, find for each connection the value of L.

[(a) 16.78 mH; (b) 73.84 mH]

7 Determine the total power dissipated by three 20 Ω resistors when connected (a) in star and (b) in delta to a 440 V, 3-phase supply.

[(a) 9.68 kW; (b) 29.04 kW]

8 Determine the power dissipated in the circuit of *Problem 3*. [1.35 kW]

9 A balanced delta connected load has a line voltage of 400 V, a line current of 8 A and a lagging power factor of 0.94. Draw a complete phasor diagram of the load. What is the total power dissipated by the load? [5.21 kW]

10 A 3-phase, star-connected alternator delivers a line current of 65 A to a balanced delta-connected load at a line voltage of 380 V. Calculate (a) the phase voltage of the alternator, (b) the alternator phase current and (c) the load phase current.

[(a) 219.4 V; (b) 65 A; (c) 37.53 A.]

11 Three inductive loads, each of resistance 4 Ω and reactance 9 Ω are connected in delta. When connected to a 3-phase supply the loads consume 1.2 kW. Calculate (a) the power factor of the load; (b) the phase current; (c) the line current and (d) the supply voltage.

[(a) 0.406; (b) 10 A; (c) 17.32 A; (d) 98.49 V.]

12 The input voltage, current and power to a motor is measured as 415 V, 16.4 A and 6 kW respectively. Determine the power factor of the system. [0.509]

13 A 440 V, 3-phase a.c. motor has a power output of 11.25 kW and operates at a power factor of 0.8 lagging and with an efficiency of 84%. If the motor is delta connected determine (a) the power input; (b) the line current and (c) the phase current. [(a) 13.39 kW; (b) 21.96 A; (c) 12.68 A.]

14 Two wattmeters are connected to measure the input power to a balanced 3-phase load. If the wattmeter readings are 9.3 kW and 5.4 kW determine (a) the total output power; and (b) the load power factor.

[(a) 14.7 kW; (b) 0.909.]

15 8 kW is found by the two-wattmeter method to be the power input to a 3-phase motor. Determine the reading of each wattmeter if the power factor of the system is 0.85. [5.431 kW; 2.569 kW]

16 Show that the power in a three phase balanced system can be measured by two wattmeters and deduce an expression for the total power in terms of the watt-meter readings. When the two-wattmeter method is used to measure the input power of a balanced load, the readings on the wattmeters are 7.5 kW and 2.5 kW, the connections to one of the coils on the meter reading 2.5 kW having to be reversed. Determine (a) the total input power and (b) the load power factor. [(a) 5 kW; (b) 0.277.]

17 Three similar coils, each having a resistance of 4.0 Ω and an inductive reactance of 3.46 Ω are connected (a) in star and (b) in delta across a 400 V, 3-phase supply. Calculate for each connection the readings on each of two wattmeters connected to measure the power by the two-wattmeter method.

[(a) 17.15 kW, 5.73 kW; (b) 51.46 kW, 17.18 kW.]

18 A 3-phase, star-connected alternator supplies a delta connected load, each phase of which has a resistance of 15 Ω and inductive reactance 20 Ω. If the line voltage is 400 V, calculate (a) the current supplied by the alternator and (b) the output power and kVA rating of the alternator, neglecting any losses in the line between the alternator and the load.   [(a) 27.71 A; (b) 11.52 kW; 19.2 kVA]

19 Each phase of a delta connected load comprises a resistance of 40 Ω and a 40 μF capacitor in series. Determine, when connected to a 415 V, 50 Hz, 3-phase supply (a) the phase current; (b) the line current; (c) the total power dissipated and (d) the kVA rating of the load.

[(a) 4.66 A; (b) 8.07 A; (c) 2.605 kW; (d) 5.80 kVA.]

20 (a) State the advantages of three-phase supplies.
   (b) Three 24 μF capacitors are connected in star across a 400 V, 50 Hz, 3-phase supply. What value of capacitance must be connected in delta in order to take the same line current?   [8 μF]

7 Transformers

A. MAIN POINTS CONCERNED WITH TRANSFORMERS

1 **Mutual inductance,** M is the property whereby an e.m.f. is induced in a circuit by a change of flux due to current changing in an adjacent circuit.

2 (i) A **transformer** is a device which uses the phenomenon of mutual induction to change the values of alternating voltages and currents. In fact, one of the main advantages of a.c. transmission and distribution is the ease with which an alternating voltage can be increased or decreased by transformers.

   (ii) Losses in transformers are generally low and thus efficiency is high. Being static they have a long life and are very reliable.

   (iii) Transformers range in size from the miniature units used in electronic applications to the large power transformers used in power stations. The principle of operation is the same for each.

3 A transformer is simply a system having an input and an output. A transformer is represented in *Fig 1(a)* as consisting of two electrical circuits linked by a common

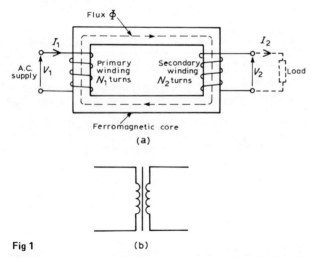

Fig 1
86

ferromagnetic core. One coil is termed the **primary winding** which is connected to the supply of electricity, and the other the **secondary winding**, which may be connected to a load. A circuit diagram symbol for the transformer is shown in *Fig 1(b)*.

4 **Transformer principle of operation.**

(i) When the secondary is an open-circuit and an alternating voltage V_1 is applied to the primary winding, a small current I_o called the no-load current, flows, which sets up a magnetic flux in the core. This alternating flux links with both primary and secondary coils and induces in them e.m.f.s of E_1 and E_2 volts respectively, by mutual induction.

(ii) The induced e.m.f E in a coil of N turns is given by $E = N\left(\dfrac{\Delta\Phi}{t}\right)$ volts, where $\left(\dfrac{\Delta\Phi}{t}\right)$ is the rate of change of flux. In an ideal transformer, the rate of change of flux is the same for both primary and secondary and thus $\dfrac{E_1}{N_1} = \dfrac{E_2}{N_2}$, i.e. **the induced e.m.f. per turn is constant.**

Assuming no losses, $E_1 = V_1$ and $E_2 = V_2$.

$$\text{Hence} \quad \frac{V_1}{N_1} = \frac{V_2}{N_2} \quad \text{or} \quad \frac{V_1}{V_2} = \frac{N_1}{N_2} \qquad (1)$$

(iii) $\dfrac{V_1}{V_2}$ is called the **voltage ratio** and $\dfrac{N_1}{N_2}$ the **turns ratio**, or the **transformation ratio** of the transformer.

If N_2 is less than N_1 then V_2 is less than V_1 and the device is termed a **step-down transformer.**

If N_2 is greater than N_1, then V_2 is greater than V_1 and the device is termed a **step-up transformer.**

(iv) When a load is connected across the secondary winding, a current I_2 flows. In an ideal transformer losses are neglected and a transformer is considered to be 100% efficient. Hence the

input power = output power

i.e., $V_1 I_1 \cos\phi_1 = V_2 I_2 \cos\phi_2$

However, the primary and secondary power factors $\cos\phi_1$ and $\cos\phi_2$ are nearly equal at full load. Hence, $V_1 I_1 = V_2 I_2$, i.e., in an ideal transformer, **the primary and secondary ampere-turns are equal.**

$$\text{Thus} \quad \frac{V_1}{V_2} = \frac{I_2}{I_1} \qquad (2)$$

Combining equations (1) and (2) gives: $\dfrac{V_1}{V_2} = \dfrac{N_1}{N_2} = \dfrac{I_2}{I_1} \qquad (3)$

5 The **rating** of a transformer is stated in terms of the **volt-amperes** that it can transform without overheating. With reference to *Fig 1(a)*, the transformer rating is either $V_1\,I_1$ or $V_2\,I_2$, where I_2 is the full-load secondary current. (See *Problems 1 to 6*.)

6 There are broadly two sources of **losses in transformers** on load — copper losses and iron losses.

(a) **Copper losses** are variable and result in a heating of the conductors, due to the fact that they possess resistance. If R_1 and R_2 are the primary and secondary winding resistances then the total copper loss is $I_1{}^2\,R_1 + I_2{}^2\,R_2$.

(b) **Iron losses** are constant for a given value of frequency and flux density and are of two types — hysteresis loss and eddy current loss.

(i) **Hysteresis loss** is the heating of the core as a result of the internal molecular structure reversals which occur as the magnetic flux alternates. The loss is proportional to the area of the hysteresis loop and thus low-loss nickel iron alloys are used for the core since their hysteresis loops have small areas.

(ii) **Eddy current loss** is the heating of the core due to e.m.f.s being induced not only in the transformer windings but also in the core. These induced e.m.f.s set up circulating currents, called eddy currents. Owing to the low resistance of the core, eddy currents can be quite considerable and can cause a large power loss and excessive heating of the core. Eddy current losses can be reduced by increasing the resistivity of the core material or, more usually, by laminating the core (i.e. splitting it into layers or leaves) when very thin layers of insulating material can be inserted between each pair of laminations. This increases the resistance of the eddy current path, and reduces the value of the eddy current.

7 **Transformer efficiency**, $\eta = \dfrac{\text{output power}}{\text{input power}} = \dfrac{\text{input power} - \text{losses}}{\text{input power}}$

$$= 1 - \frac{\text{losses}}{\text{input power}}$$

and is usually expressed as a percentage. It is not uncommon for power transformers to have efficiencies of between 95% and 98%. Output power = $V_2 I_2 \cos \phi_2$, losses = copper loss + iron losses, and input power = output power + losses. (See *Problems 7 and 8*)

8 **Transformer construction.**

(i) There are broadly two types of transformer construction — the **core type** and the **shell type**, as shown in *Fig 2*. The low and high voltage windings are wound as shown, to reduce leakage flux.

(ii) For **power transformers**, rated possibly at several MVA and operating at a frequency of 50 Hz in Great Britain, the core material is usually laminated

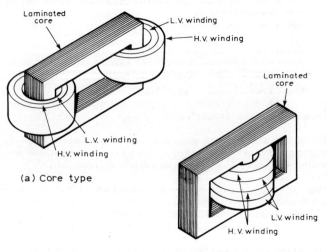

(a) Core type

(b) Shell type

Fig 2

silicon steel or stalloy, the laminations reducing eddy currents and the silicon steel keeping hysteresis loss to a minimum. Large power transformers are used in the main distribution system and in industrial supply circuits. Small power transformers have many applications, examples including welding and rectifier supplies, domestic bell circuits, imported washing machines, and so on.

(iii) For **audio-frequency (a.f.) transformers**, rated from a few mVA to no more than 20 VA, and operating at frequencies up to about 15 kHz, the small core is also made of laminated silicon steel. A typical application of a.f. transformers is in an audio amplifier systems.

(iv) **Radio-frequency (r.f.) transformers**, operating in the MHz frequency region have either an air core, a ferrite core or a dust core. Ferrite is a ceramic material having magnetic properties similar to silicon steel, but having a high resistivity. Dust cores consist of fine particles of carbonyl iron or permalloy (i.e. nickel and iron), each particle of which is insulated from its neighbour. Applications of r.f. transformers are found in radio and television receivers.

(v) Transformer **windings** are usually of enamel-insulated copper or aluminium.

(vi) **Cooling** is achieved by air in small transformers and oil in large transformers.

9 Transformers not only enable current or voltage to be transformed to some different magnitude but provide a means of isolating electrically one part of a circuit from another when there is no electrical connection between primary and secondary windings. An **isolating transformer** is a 1 : 1 ratio transformer with several important applications, including bathroom shaver-sockets, portable electric tools, model railways, and so on.

B. WORKED PROBLEMS ON TRANSFORMERS

Problem 1 A transformer has a primary winding of 100 turns and a secondary winding of 40 turns. If the primary voltage is 240 V determine the secondary voltage for an ideal transformer.

Voltage ratio = turns ratio, or $\dfrac{V_1}{V_2} = \dfrac{N_1}{N_2}$

Hence secondary voltage $V_2 = V_1\left(\dfrac{N_2}{N_1}\right) = 240\left(\dfrac{40}{100}\right) = $ **96 volts**

Problem 2 An ideal transformer with a turns ratio of 2 : 5 is fed from a 200 V supply. Determine its output voltage.

A turns ratio of 2 : 5 means that the transformer has 2 turns on the primary for every 5 turns on the secondary, i.e. $\dfrac{N_1}{N_2} = \dfrac{2}{5}$. This is a step-up transformer.

$\dfrac{N_1}{N_2} = \dfrac{V_1}{V_2}$, from which secondary voltage $V_2 = V_1\left(\dfrac{N_2}{N_1}\right) = 200\left(\dfrac{5}{2}\right)$
$= $ **500 volts**

Problem 3 An ideal transformer has a turns ratio of 8 : 1 and the primary current is 3 A when it is supplied at 240 V. Calculate the secondary voltage and current.

A turns ratio of 8 : 1 means $\dfrac{N_1}{N_2} = \dfrac{8}{1}$, thus a step-down transformer.

$\dfrac{N_1}{N_2} = \dfrac{V_1}{V_2}$ from which secondary voltage $V_2 = V_1\left(\dfrac{N_2}{N_1}\right) = 240\left(\dfrac{1}{8}\right) = \textbf{30 volts}$

Also, $\dfrac{N_1}{N_2} = \dfrac{I_2}{I_1}$. Hence secondary current, $I_2 = I_1\left(\dfrac{N_1}{N_2}\right) = 3\left(\dfrac{8}{1}\right) = \textbf{24 A}$

Problem 4 An ideal transformer, connected to a 240 V mains, supplies a 12 V, 150 W lamp. Calculate the transformer turns ratio and the current taken from the supply.

$V_1 = 240$ V, $V_2 = 12$ V; $I_2 = \dfrac{P}{V_2} = \dfrac{150}{12} = 12.5$ A

Turns ratio $= \dfrac{N_1}{N_2} = \dfrac{V_1}{V_2} = \dfrac{240}{12} = \textbf{20}$

$\dfrac{V_1}{V_2} = \dfrac{I_2}{I_1}$, from which $I_1 = I_2\left(\dfrac{V_2}{V_1}\right) = 12.5\left(\dfrac{12}{240}\right)$

Hence current taken from the supply, $I_1 = \dfrac{12.5}{20} = \textbf{0.625 A}$

Problem 5 A 12Ω resistor is connected across the secondary winding of an ideal transformer whose secondary voltage is 120 V. Determine the primary voltage if the supply current is 4 A.

Secondary current $I_2 = \dfrac{V_2}{R_2} = \dfrac{120}{12} = 10$ A

$\dfrac{V_1}{V_2} = \dfrac{I_2}{I_1}$, from which the primary voltage $V_1 = V_2\left(\dfrac{I_2}{I_1}\right) = 120\left(\dfrac{10}{4}\right) = \textbf{300 volts}$

Problem 6 A 5 kVA single-phase transformer has a turns ratio of 10 : 1 and is fed from a 2.5 kV supply. Neglecting losses, determine (a) the full-load secondary current, (b) the minimum load resistance which can be connected across the secondary winding to give full load kVA, (c) the primary current at full load kVA.

(a) $\dfrac{N_1}{N_2} = \dfrac{10}{1}$; $V_1 = 2.5$ kV $= 2500$ V.

Since $\dfrac{N_1}{N_2} = \dfrac{V_1}{V_2}$, secondary voltage $V_2 = V_1\left(\dfrac{N_2}{N_1}\right) = 2500\left(\dfrac{1}{10}\right) = 250$ V.

The transformer rating in volt-amperes $= V_2 I_2$ (at full load),
i.e. $5000 = 250 I_2$

Hence full load secondary current $I_2 = \dfrac{5000}{250} = \textbf{20 A}$

(b) Minimum value of load resistance, $R_L = \dfrac{V_2}{I_2} = \dfrac{250}{20} = \textbf{12.5}\Omega$

(c) $\dfrac{N_1}{N_2} = \dfrac{I_2}{I_1}$, from which primary current $I_1 = I_2\left(\dfrac{N_2}{N_1}\right) = 20\left(\dfrac{1}{10}\right) = \textbf{2A}$

Efficiency $\eta = \dfrac{\text{output power}}{\text{input power}} = \dfrac{\text{input power} - \text{losses}}{\text{input power}} = 1 - \dfrac{\text{losses}}{\text{input power}}$

Full-load output power = $V I \cos \phi = (200)(0.85) = 170$ kW
Total losses = 1.5 + 1.0 = 2.5 kW
Input power = output power + losses = 170 + 2.5 = 172.5 kW
Hence efficiency = $1 - \dfrac{2.5}{172.5} = 1 - 0.01449 = 0.9855$ or **98.55%**

Half full-load power output = $\dfrac{1}{2}$ (200)(0.85) = 85 kW

Copper loss (or $I^2 R$ loss) is proportional to current squared. Hence the copper loss at half full-load is $\left(\dfrac{1}{2}\right)^2$ (1500) = 375 W

Iron loss = 1000 W (constant)
Total losses = 375 + 1000 = 1375 W or 1.375 kW
Input power at half full-load = output power at half full-load + losses
$\qquad\qquad\qquad\qquad = 85 + 1.375 = 86.375$ kW
Hence efficiency = $1 - \dfrac{\text{losses}}{\text{input power}} = 1 - \dfrac{1.375}{86.375} = 1 - 0.01592 = 0.9841$

$\qquad\qquad\qquad\qquad\qquad\qquad\qquad\qquad\qquad$ or **98.41%**

C. FURTHER PROBLEMS ON TRANSFORMERS

(a) SHORT ANSWER PROBLEMS

1 Define mutual inductance.
2 Describe briefly the principle of operation of a transformer.
3 How is a transformer rated?
4 Name the two main sources of loss in a transformer.
5 What is hysteresis loss? How is it minimised in a transformer?
6 What are eddy currents? How may they be reduced in transformers?
7 What is an isolating transformer? Give two applications.
8 Name two types of transformer construction.
9 What core material is used normally for power transformers?
10 Name three core materials used in r.f. transformers.

(b) MULTI-CHOICE PROBLEMS (answers on page 125)

1 An ideal transformer has a turns ratio of 1 : 5 and is supplied at 200 V when the primary current is 3 A.

Which of the following statements is false?
(a) The turns ratio indicates a step-up transformer.
(b) The secondary voltage is 40 V.
(c) The secondary current is 15 A.
(d) The transformer rating is 0.6 kVA.
(e) The secondary voltage is 1 kV.
(f) The secondary current is 0.6 A.

A 100 kVA, 250 V/10 kV, single-phase transformer has a full-load copper loss of 800 W and an iron loss of 500 W. The primary winding contains 120 turns. For the statements in *Problems 2 to 8*, select the correct answer from this list.

(a) 81.3 kW   (b) 800 W   (c) 97.32%   (d) 80 kW   (e) 3
(f) 4800   (g) 1.3 kW   (h) 98.40%   (i) 100 kW   (j) 98.28%
(k) 200 W   (l) 101.3 kW (m) 96.38%   (n) 400 W

2   The total full-load losses.
3   The full-load output power at 0.8 power factor.
4   The full-load input power at 0.8 power factor.
5   The full-load efficiency at 0.8 power factor.
6   The half full-load copper loss.
7   The transformer efficiency at half full-load, 0.8 power factor.
8   The number of secondary winding turns.

9   Which of the following statements is false?
    (a) In an ideal transformer, the volts per turn are constant for a given value of primary voltage.
    (b) In a single-phase transformer, the hysteresis loss is proportional to frequency.
    (c) A transformer whose secondary current is greater than the primary current is a step-up transformer.
    (d) In transformers, eddy-current loss is reduced by laminating the core.

(c) CONVENTIONAL PROBLEMS

1   A transformer has 600 primary turns connected to a 1.5 kV supply. Determine the number of secondary turns for a 240 V output voltage, assuming no losses.

[96]

2   An ideal transformer with a turns ratio of 2 : 9 is fed from a 220 V supply. Determine its output voltage.

[990 V]

3   (a) Describe the transformer principle of operations.
    (b) An ideal transformer has a turns ratio of 12 : 1 and is supplied at 180 V when the primary current is 4 A. Calculate the secondary voltage and current.

[15 V; 48 A]

4   A step-down transformer with a turns ratio of 20 : 1 has a primary voltage of 4 kV and a load of 10 kW. Neglecting losses, calculate the value of the secondary current.

[50 A]

5   A transformer has a primary-to-secondary turns ratio of 1 : 15. Calculate the primary voltage necessary to supply a 240 V load. If the load current is 3 A, determine the primary current. Neglect any losses.

[16 V; 45 A]

6   A 10 kVA, single-phase transformer has a turns ratio of 12 : 1 and is supplied from a 2.4 kV supply. Neglecting losses, determine (a) the full load secondary current,

(b) the minimum value of load resistance which can be connected across the secondary winding without the kVA rating being exceeded, and (c) the primary current.

[(a) 50 A (b) 4Ω (c) 4.17 A]

7   A 500/100 V, single-phase transformer takes a full-load primary current of 4 A. Neglecting losses, determine (a) the full-load secondary current, (b) the rating of the transformer.

[(a) 20 A (b) 2 kVA]

8   State the essential features of construction of power, a.f. and r.f. transformers and state one application of each type. A 20Ω resistance is connected across the secondary winding of a single-phase power transformer whose secondary voltage is 150 V. Calculate the primary voltage and the turns ratio if the supply current is 5 A; neglecting losses.

[225 V; 3 : 2]

9   A single-phase transformer has a voltage ratio of 6 : 1 and the h.v. winding is supplied at 540 V. The secondary winding provides a full load current of 30 A at a power factor of 0.8 lagging. Neglecting losses, find (a) the rating of the transformer, (b) the power supplied to the load, (c) the primary current.

[(a) 2.7 kVA, (b) 2.16 kW, (c) 5A]

10 A single-phase transformer is rated at 40 kVA. The transformer has full-load copper losses of 800 W and iron losses of 500 W. Determine the transformer efficiency at full load and 0.8 power factor.

[96.10%]

11 Determine the efficiency of the transformer in *Problem 10* at half full-load and 0.8 power factor.

[95.81%]

12 (a) Explain the choice of transformer core materials and construction to minimise core losses.
    (b) A 100 kVA, 2000/400 V, 50 Hz, single-phase transformer has an iron loss of 600 W and a full-load copper loss of 1600 W. Calculate its efficiency for a load of 60 kW at 0.8 power factor.

[97.56%]

13 (a) What are eddy currents? State how their effect is reduced in transformers.
    (b) Determine the efficiency of a 15 kVA transformer for the following conditions:
        (i) full-load, unity power factor
        (ii) 0.8 full-load, unity power factor,
        (iii) half full-load, 0.8 power factor.
    Assume that iron losses are 200 W and the full-load copper loss is 300 W.

[(a) 96.77% (ii) 96.84% (iii) 95.62%]

8 Electronic systems

A. MAIN POINTS CONCERNED WITH ELECTRONIC SYSTEMS

1   A **system** is a group of components connected together to perform a desired function. *Fig 1* illustrates a simple hi-fi system.

2   A **sub-system** is a part of a system which performs an identified function within a whole system. For example, the pre-amplifier/tone control/input switching section of *Fig 1* is a conveniently identified sub-system. Its function is to manipulate input signals (select or mix them and modify their base and treble content) prior to the main amplification.

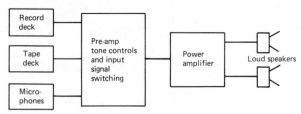

Fig 1

3   A **component or element** is usually the simplest part of a system which has a specific and well-defined function; for example, a microphone in *Fig 1*.

4   An **electronic amplifier stage** is an active amplifying component together with the **coupling components** to the next stage. A stage begins with the input to an amplifying component (e.g. a transistor) and ends at the input to the next amplifying component. *Fig 2* illustrates this point. Transistor T_1 is the amplifying component and the components between T_1 and T_2 are the coupling components. Each transistor has three connection points (terminals) these being the **collector** (marked 'c'), the **base** (marked 'b') and the **emitter** (marked 'e'). For the amplifier to operate satisfactorily, certain fixed direct voltage (d.c.) levels are maintained across the components of a stage. An alternating (a.c.) signal is fed to the base of T_1, which is amplified by T_1 and appears as an amplified version of the a.c. signal at the base of T_2. For the amplifier shown in *Fig 2*, a small a.c. voltage V_1 between the base and the zero line causes a small a.c. current to flow into the base. The size of this current depends on the **input resistance** R_{IN} of the transistor. The transistor amplifies this input current so that a much larger a.c. current flows from the

emitter to the collector out into the components connected to the collector. With good design most of this collector a.c. current, which is an amplified version of the original signal current, can be made to flow into the base of the next transistor. In this way successive stages amplify the original signal current. The capacitors are only present to prevent d.c. conditions in various parts of the circuit interfering with each other. Voltages representing amplified versions

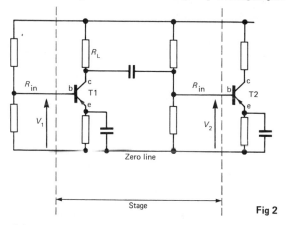

Fig 2

of the original a.c. signal voltage appear at the collector of T_1 and the base of T_2, and then at the collector of T_2.

Note that the stage includes the input resistance to the next part of the circuit. This may be another amplifier stage or a load such as a loudspeaker or a transformer.

Fig 3 shows this concept where the triangle represents an amplifying stage.

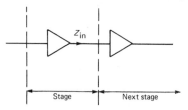

Fig 3

5 The **stage gain** is the ratio of output quantity to input quantity for a stage. When the input and output quantities are of the same kind the stage gain is dimensionless. In electronic circuits, stage gain G depends on the frequency of the signal and this results in a phase difference ϕ between output and input signals. In *Fig 2*,

Stage gain $= \dfrac{V_2 \angle \phi}{V_1} = G \angle \phi$

where $\angle \phi$ varies with the frequency of the input signal. In addition, when a transistor is the amplifying component, there is a 180° phase difference between the base and the collector signals, i.e. the output voltage is 180° out of phase with the input.

95

6   The **overall gain** of an electronic amplifier having several stages is given by $G = G_1 \times G_2 \times \ldots\ldots \times G_n$, i.e. the **product** of the stage gains. Suffixes V, A and P are used to distinguish voltage, current and power gains respectively.

Thus $G_V = G_{V_1} \times G_{V_2} \times \ldots\ldots \times G_{V_n}$

$\quad\quad G_A = G_{A_1} \times G_{A_2} \times \ldots\ldots \times G_{A_n}$

$\quad\quad G_P = G_{P_1} \times G_{P_2} \times \ldots\ldots \times G_{P_n}$

Further, as the overall voltage gain $G_V = \dfrac{V_{OUT}}{V_{IN}}$

and the overall curent gain $G_A = \dfrac{I_{OUT}}{I_{IN}}$

then $G_V \times G_A = \dfrac{V_{OUT}}{V_{IN}} \times \dfrac{I_{OUT}}{I_{IN}}$. But power $P = VI$,

hence $G_V \times G_A = \dfrac{P_{OUT}}{P_{IN}}$

Thus $\quad G_P = G_V \times G_A$

This applies for both the amplifier as a whole and for each individual stage.

7   In an electronic amplifier, the **interstage coupling** network is an arrangement of resistors and capacitors (and sometimes inductors or transformers) which allows an a.c. signal to proceed from one stage to the next, but prevents the fixed d.c. conditions of one stage from affecting those of the next. *Fig 4* shows a typical arrangement. The capacitor C prevents the d.c. collector potential V_{C_1} from affecting the d.c. base potential V_{B_2}.

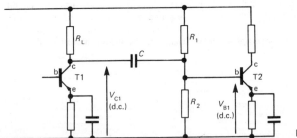

Fig 4

At high frequencies, capacitor C is virtually a short circuit $X_C = 1/2\pi fC$, so that the whole of the a.c. signal voltage at T_1 collector is passed to the base of T_2. At lower frequencies, the capacitor C and resistors R_1 and R_2 form a potential dividing network which reduces the a.c. signal voltage at the base of T_2.

B. WORKED PROBLEMS ON ELECTRONIC SYSTEMS

Problem 1 A system is required to have a voltage gain of 1400. If two stages are to provide this and one stage has a stage gain of twice that of the other stage, calculate the individual stage gains.

From para. 6, $G_V = G_{V_1} \times G_{V_2}$

But $G_{V_1} = 2 G_{V_2}$

Therefore $G_V = 2 G_{V_2} \times G_{V_2} - 2(G_{V_2})^2$

Thus $G_{V_2} = \sqrt{\dfrac{G_V}{2}}$

But $G_V = 1400$, hence

$$G_{V_2} = \sqrt{\frac{1400}{2}} = \sqrt{700}$$
$$= 26.46$$

As $G_{V_1} = 2 G_{V_2}$,

$$G_{V_1} = 2 \times 26.46$$
$$= 52.92$$

Thus the individual stage gains are 52.92 and 26.46

Problem 2 An electronic amplifier has an input signal of 25 μV and delivers 3 W into an 8Ω load. What is the overall voltage gain?

The power $P = \dfrac{V^2}{R}$ watts

Let the load voltage be V_L volts, then

$$3 = \frac{V_L^2}{8}$$

and $V_L = \sqrt{24}$

From para 6, the overall voltage gain $G_V = \dfrac{V_{OUT}}{V_{IN}} = \dfrac{V_L}{V_{IN}}$

$$= \frac{\sqrt{24}}{25 \times 10^{-6}}$$
$$= 196 \times 10^3$$

Problem 3 An amplifier consists of three stages having identical voltage gains of 85. If the maximum peak-to-peak output voltage V_0 is 28 V, calculate the maximum r.m.s. input voltage V_{IN}.

Since the ratio $\dfrac{\text{peak value}}{\text{r.m.s. value}}$ of an a.c. waveform is $\sqrt{2}$ and since the peak value is half the peak-to-peak value, then

V_{OUT} r.m.s. $= \dfrac{28}{2\sqrt{2}} = 9.9$ V

As the overall gain $G_V = G_{V_1} \times G_{V_2} \times G_{V_3}$ and $G_{V_1} = G_{V_2} = G_{V_3} = 85$, $G_V = 85^3$

But $G_V = \dfrac{V_{OUT}}{V_{IN}}$, thus V_{IN} r.m.s $= \dfrac{9.9}{(85)^3}$

$$= 16.12 \ \mu\text{V}$$

Problem 4 A two-stage amplifier has stage gains at a certain frequency:
$G_{V_1} = 37\angle{-80^\circ}$, $G_{V_2} = 12\angle{-75^\circ}$
Calculate the overall voltage gain and the phase difference between output and input voltages.

From para. 6, overall gain is the product of individual gains. In general,
if $G_{V_1} = r_1 \angle \theta_1$ and $G_{V_2} = r_2 \angle \theta_2$, then $G_{V_1} \times G_{V_2} = (r_1 \times r_2) \angle (\theta_1 + \theta_2)$.
Similarly, $G_{V_1} \times G_{V_2} \times G_{V_3} = (r_1 \times r_2 \times r_3) \angle (\theta_1 + \theta_2 + \theta_3)$.
Thus $G_V = (37 \angle -80°) \times (12 \angle -75°)$
$= (37 \times 12) \angle (-80° - 75°)$
$= 444 \angle -155°$
$= \mathbf{444 \angle +205°}$

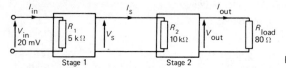

Fig 5

Problem 5 A two-stage amplifier, shown in *Fig 5*, has an input of 20 mV. Stage 1 has an input resistance of 5 kΩ and a voltage gain of 20. Stage 2 has an input resistance of 10 kΩ and a voltage gain of 40. For a load of 80Ω, calculate (a) the input current I_s to the second stage, (b) the overall voltage gain, (c) the overall current gain, (d) the overall power gain.

The solution of problems such as this relies largely on the use of Ohm's law, i.e.
$V = I R$, and the gain relationships introduced in para 6,

i.e., $G = \dfrac{\text{output quantity}}{\text{input quantity}}$

For stage 1 input:
$$I_{IN} = \frac{V_{IN}}{R_1} = \frac{20 \text{ mV}}{5 \text{ k}\Omega} = \frac{20 \times 10^{-3}}{5 \times 10^3}$$
$$= 4 \times 10^{-6} \text{A} \qquad = 4 \text{ }\mu\text{A}$$

For stage 2 input:
$$G_{V_1} = \frac{V_S}{V_{IN}}, \text{ so } V_S = G_{V_1} \times V_{IN}$$
$$= 20 \times (20 \times 10^{-3}) = 4 \times 10^{-1} = 0.4 \text{ V}$$

(a) $I_S = \dfrac{V_S}{R_2} = \dfrac{0.4}{10 \times 10^3}$
$$= 4 \times 10^{-5} = \mathbf{40 } \text{ } \boldsymbol{\mu}\mathbf{A}$$

For stage 2 output:
$$V_{OUT} = G_{V_2} \times V_{IN} = 40 \times 0.4 = 16 \text{ V}$$
$$I_{OUT} = \frac{V_{OUT}}{R_{LOAD}} = \frac{16}{80} = 0.2 \text{ A}$$

(b) The overall voltage gain, $G_V = \dfrac{V_{OUT}}{V_{IN}} = \dfrac{16}{20 \times 10^{-3}} = \mathbf{800}$

[Check: $G_V = G_{V_1} \times G_{V_2} = 20 \times 40 = 800$]

(c) The overall current gain, $G_A = \dfrac{I_{OUT}}{I_{IN}} = \dfrac{0.2 \text{ A}}{4 \text{ } \mu\text{A}}$
$$= \frac{0.2}{4 \times 10^{-6}} = \mathbf{50\,000}$$

(d) The overall power gain, $G_P = G_V \times G_A$
$$= 800 \times 50\,000 = \mathbf{40 \times 10^6}$$

$$\left[\text{Check: } G_p = \frac{V_{OUT}\, I_{OUT}}{V_{IN}\, I_{IN}} = \frac{16 \times 0.2}{20 \times 10^{-3} \times 4 \times 10^{-6}}\right.$$
$$\left. = 40 \times 10^6 \right]$$

Problem 6 A transistor amplifier has three stages with interstage networks giving 10°, 12° and 15° phase advances respectively at a particular frequency. Allowing for a phase advance of 180° for each transistor in addition to the phase advance for frequency, determine the phase angle of the output relative to the input.

At the output of the first stage, shown as (1) in *Fig 6*, there is a phase shift of 180° due to the transistor and 10° due to the interstage coupling, giving a total phase advance of 190° for this stage. Similarly, for stage 2 the total phase advance is 180° + 12° = 192°. For stage 3, the total phase advance is 180° + 15° = 195°.

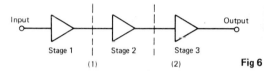

Input   Stage 1   Stage 2   Stage 3   Output

(1)   (2)   **Fig 6**

The phase advance at the output is the sum of the three phase advances associated with the three stages, 190° + 192° + 195° = 577°.
It is usual to express a phase advance in terms of an angle between 0° and 360°, and since 577° = 577° − 360°, then the phase advance of the amplifier output relative to the input is (577 − 360)° = 217°.

C. FURTHER PROBLEMS ON ELECTRONIC SYSTEMS

(a) SHORT ANSWER PROBLEMS

1   A system is a group of ...... connected so as to perform a ......
2   A sub-system is a ...... of a system identified by its ......
3   An amplifier stage contains the active ...... and the ...... components to the next stage.
4   Stage gain is defined as the ratio of ...... to ......
5   If the output from a system is measured in amperes and the input is measured in millivolts, the dimensions of the overall gain is ....
6   The power gain of an electronic system is the product of .... and ....
7   A .... is used in interstage coupling to prevent d.c. conditions from interacting between stages.

(b) MULTI-CHOICE PROBLEMS (answers on page 125)

Select the correct answer to the following problems.
1   A system is designed to:
  (a) be divided into sub-systems
  (b) produce an output proportional to input
  (c) produce an output in response to an input
  (d) produce a constant output

2 The gain of a system is:
   (a) dimensionless
   (b) the ratio of input to output
   (c) the ratio of output to input
   (d) the sum of the stage gains
3 A hi-fi system amplifies:
   (a) a record groove
   (b) a magnetic tape
   (c) sound
   (d) electrical signals
4 In a transistor amplifier having two stages, the d.c. bias conditions are:
   (a) kept separate by the interstage capacitor
   (b) reduced by the interstage network
   (c) amplified by the system
   (d) applicable only to the collector
5 An amplifier stage of a transistor amplifier is defined as:
   (a) all the components from the base of one transistor to the base of the next
   (b) all the components from the base of a transistor to the collector
   (c) the network between the collector of one transistor and the base of the next
   (d) the transistor alone.
6 An amplifier with three stages having identical stage gains G has overall gain G_V
   where:
   (a) $G_V = 3 \times (G)$
   (b) $G_V = (G)^3$
   (c) $G_V = (G)^{\frac{1}{3}}$
   (d) $G_V = 3 + G$

(c) CONVENTIONAL PROBLEMS

1 The stages of a three-stage amplifier have voltage gains of 23, 18 and 13. Calculate
   the overall voltage gain.

   [5382]

2 A four-stage system has an overall gain of 20 000. Two of the stages have gains of
   20 and 15 and the other two have equal stage gains. Calculate the stage gains of
   the other stages.

   [8.16]

3 An amplifier has a power gain of 800. Calculate the input current (r.m.s.) for an
   8 W output if the input resistance is 1 kΩ.

   [3.162 mA]

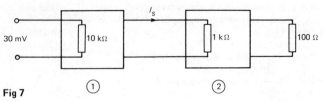

Fig 7

4 The two-stage amplifier in *Fig 7* is driven by a 30 mV input signal. Stage 1 has a
   voltage gain of 40 and an input resistance of 10 kΩ. Stage 2 has a voltage gain of
   60 and an input resistance of 1 kΩ. If the load is 100Ω, calculate:

100

(a) the overall voltage gain
(b) the overall current gain
(c) the overall power gain
(d) the interstage current I_s.

[(a) 2400; (b) 240 000; (c) 576 × 10^6; (d) 1.2 mA]

5 The two stages of a system have voltage gains of $30\angle25°$ and $42\angle28°$ at a particular frequency. Calculate the overall voltage gain at this frequency.

[$1260\angle53°$]

6 A transistor amplifier consists of three stages, with interstage networks giving 20°, 22° and 28° phase advances respectively. Taking into account the 180° phase shifts of each transistor, determine the overall phase angle between input and output.

[250°]

9 Basic digital circuits

A. MAIN POINTS CONCERNED WITH BASIC DIGITAL ELECTRONIC CIRCUITS

1   A logic '1' is a predetermined potential, different from zero. It may be a positive voltage or a negative voltage depending on the circuit design.

2   A logic '0' is zero electric potential.

3   An AND gate is an electronic circuit having several inputs and one output, in which a logical 1 appears at the output if all inputs are simultaneously at logic 1 level.

4   A NAND gate is an electronic circuit whose output is a logical 0 if all inputs are at logical 1 level.

5   An OR gate is an electronic circuit whose output is a logical 1 if any of its inputs is at logical 1 level.

6   A NOR gate is an electronic circuit whose output is a logical 0 if any of its inputs is a logical 1.

7   A NOT gate is an electronic circuit with one input and one output whose output adopts the opposite logical sense to its input.

8   *Fig 1* illustrates the basic circuits for NOR, NAND and NOT gates together with their British Standard symbols.

The basic logic gates use combinations of **diodes** and **transistors**. A diode allows current to flow in one direction only, in the direction indicated by the triangle. Thus, in *Fig 1(a)*, if point A is more positive than point X, diode D_1 allows current to pass. D_1 is then said to be **forward biased**, and in this condition point X is very nearly at the same potential as A.

The transistor operates as a switch and is either made to conduct heavily (switched on), making its collector potential almost zero, or it is not conducting at all (switched off), making its collector potential equal to $+V$. There are no 'in-between' potentials allowed for the collector potential in logic circuits.

In *Fig 1(a)*, diodes D_1 and D_2 together with R_B (the base input resistance) form an OR gate. Point X will move to logical 1 if either A or B (or both) is at logic 1. The transistor is switched on when X is at logical 1, causing C to move to a logical 0 under these conditions i.e., the transistor inverts the input. R_C ensures the transistor is OFF when A and B are both at logical 0 level. The result is a NOT-OR, i.e, a NOR operation. In *Fig 1(b)*, D_1, D_2 and R_B form an AND gate. If both A and B are at logical 1, then point X, if it were free to do so, would move to logical 1. The transistor is supplied with base current via R_B and D_3 and is switched on, thus its collector moves to logic 0 under these conditions. R_C ensures the transistor

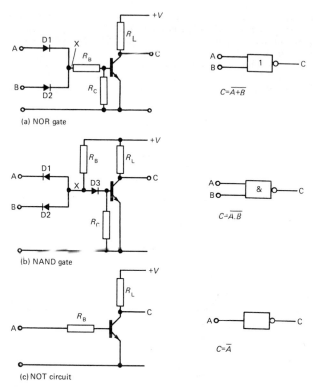

(a) NOR gate

$$C = \overline{A+B}$$

(b) NAND gate

$$C = \overline{A \cdot B}$$

(c) NOT circuit

$$C = \overline{A}$$

Fig 1

is OFF when the input is at logical 0. The result is a NOT-AND i.e. a NAND operation. The purpose of D_3 is to remove the small potential at point X when either input is zero, due to the forward voltage drop across D_1 and D_2.

In *Fig 1(c)*, if A is at logical 1 the transistor conducts and C is then at logical 0. The result is a NOT operation.

Note. (1) A NOR gate with all inputs except one connected to the zero volts line acts as a NOT gate. Similarly, a NAND gate with all inputs except one connected to the +V line acts as a NOT gate.

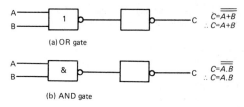

(a) OR gate

$$C = \overline{\overline{A+B}}$$
$$\therefore C = A+B$$

(b) AND gate

$$C = \overline{\overline{A \cdot B}}$$
$$\therefore C = A \cdot B$$

Fig 2

(2) The AND and OR gates can be produced by following a NAND with a NOT and a NOR with a NOT as shown in *Fig 2*. The NOT circuit inverts the output from the gate so that, for example, NOT-AND becomes NOT-NOT-AND, or simply AND.

9 A **truth table** gives the output conditions for all possible combinations of input conditions for a gate, or for a more complex logic circuit.

A truth table for the NOR, OR, NAND and AND functions having two inputs is shown below.

A	B	NOR $\overline{A + B}$	OR $A + B$	NAND $\overline{A \cdot B}$	AND $A \cdot B$
0	0	1	0	1	0
0	1	0	1	1	0
1	0	0	1	1	0
1	1	0	1	0	1

10 Binary numbers consist of zeros and ones only. For example, the number 1010 represents $1 \times 2^3 + 0 \times 2^2 + 1 \times 2^1 + 0 \times 2^0$ which in decimal is equivalent to 10. To add two binary numbers the following rules apply:

$0 + 0 = 0$
$0 + 1 = 1$
$1 + 0 = 1$
$1 + 1 = 0$ carry 1

The two numbers to be added are set out in the same way as for decimal addition, with the least significant column on the right. For example:

Decimal:	10^3	10^2	10^1	10^0		Binary:	2^4	2^3	2^2	2^1	2^0
	3	9	2	0			0	1	0	1	0
plus	1	7	1	6		plus	0	0	1	1	1
Sum	5	6	3	6		Sum	1	0	0	0	1
Carry	1	0	0			Carry	1	1	1	0	

For binary addition, the right-hand column is added first and the result is entered in the sum row and the carry (1 or 0) is placed under the next column to be added with that column.

A **binary half adder** is a combination of gates whose outputs are the sum of two binary digits and the carry resulting from this addition. The circuit and truth table are given in *Fig 3*.

From the table in *Fig 3*, S represents the sum for each combination of A and B, and C represents the carry. Thus, for example, in the second row of *Fig 3*, 0 plus 1 = sum 1 carry 0.

11 A **binary full adder** is a combination of two half adders and an OR function whose outputs represent the sum of three binary digits and the carry resulting from this addition.

Fig 4 shows the circuit and truth table for a binary full adder.

S_2 represents the sum for all conditions of A, B and C, and CO representing the carry resulting from this addition. Thus, for example, row 4 means that $A = 0$, $B = 1$ and there is a carry of 1 from the previous addition. A plus B gives a sum,

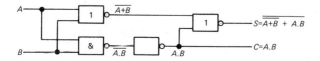

A	B	A+B	$\overline{A+B}$	$S=\overline{\overline{A+B}+A.B}$	C=A.B
0	0	0	1	0	0
0	1	1	0	1	0
1	0	1	0	1	0
1	1	1	0	0	1

Fig 3

S_1 of 1 in column 4 and carry C_1 of 0 in column 5. The column 4 result S_1 is then added to the previous carry C (column 3) to give a 0 sum S_2 in column 6, and a carry C_2 of 1 in column 7. C and C_2 are fed to an OR gate to give CO in column 8.

12 Electronic counters use **bistable elements**. A bistable is an electronic circuit which has two stable states representing binary 1 and binary zero. It can be switched from one state to the other by a pulse of voltage moving, for example, from $+V$ to 0 applied to its input. Once it has switched, it remains in that state until another pulse moving from $+V$ to 0 is applied, when it switches to the other state. The output from one bistable can be used to trigger another bistable so that, for

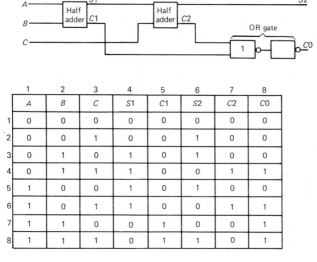

	1	2	3	4	5	6	7	8
	A	B	C	S1	C1	S2	C2	CO
1	0	0	0	0	0	0	0	0
2	0	0	1	0	0	1	0	0
3	0	1	0	1	0	1	0	0
4	0	1	1	1	0	0	1	1
5	1	0	0	1	0	1	0	0
6	1	0	1	1	0	0	1	1
7	1	1	0	0	1	0	0	1
8	1	1	1	0	1	1	0	1

Fig 4

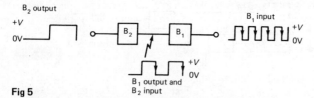

Fig 5

example, in *Fig 5*, a train of input pulses is applied to bistable B_1, and every pulse from $+V$ to 0 on the input results in B_1 output changing its state. When the output of B_1 pulses from $+V$ to 0 it triggers B_2, and B_2 output changes state. Thus the B_1 output works at half the rate of the input train of pulses and B_2 output works at half the rate of B_1 output. Thus the circuit divides successively by 2.

13 A **decade counter** is a collection of four electronic bistable (2-state) elements and gates arranged so that input pulses are stored in binary number form up to a binary count of 1001 (representing decimal 9). On the tenth input pulse the bistables all reset to zero, and a 'carry' output pulse is available.

A decoder is required to read the binary state of the bistable and present a decimal display. *Fig 6* illustrates this arrangement.

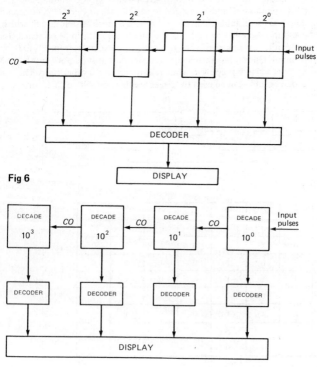

Fig 6

Fig 7

A **cascaded decade counter** is a collection of decade counters in which the input to successive counters is the carry, CO from the previous counter.

Fig 7 shows an arrangement for counting up to 9999 input pulses.

B. WORKED PROBLEMS ON BASIC DIGITAL CIRCUITS

Problem 1 Two binary digits A and B can be added to produce a binary sum S and a carry C according to the expressions

$S = (A + B) . (\overline{A . B})$

$C = A . B$

Devise a system which uses AND, OR and NOT gates to perform this function, and label the output from each stage.

From inputs A and B it is necessary to generate the Boolean functions $A + B$, $A . B$ and $\overline{A . B}$, then combine these as required by the formulae. A suitable arrangement is as shown in *Fig 8*. Note that the OR and AND gates in this figure performs

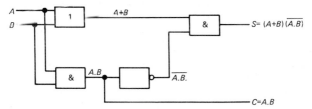

Fig 8

exactly the same logical functions as the gates shown in *Fig 2* (i.e. $C = A + B$ and $C = A . B$ respectively).

Problem 2 Explain why a half adder is not sufficient to perform addition in a computer. By reference to a block diagram, explain how a full adder overcomes the deficiencies of a half adder.

In a computer two binary numbers may be added together, in which digits of different powers of 2 are represented. Thus, for example, to add two numbers A and B, where $A = 0101$ and $B = 0011$, the addition process is as shown.

	2^3	2^2	2^1	2^0
$A =$	0	1	0	1
$B =$	0	0	1	1
$A + B =$	1	0	0	0
Carry	1	1	1	

Any of the columns except the 2^0 column will have a carry (either a 1 or a 0) from the previous column to be added in. Thus for any column three binary digits (A, B

and Carry) are to be added. The half adder is only capable of adding two digits and so is not sufficient for the above purpose.

A full adder is used to perform additions in computers and microprocessors and consists of two half adders and an OR gate as shown in *Fig 9*.

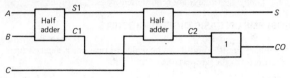

Fig 9

The sum, S_1 of digits A and B is added to digit C (the carry from the previous column), to produce the sum S.

A carry, CO, is generated if either or both the half adder carry functions result in a logical 1. Carry CO is available for consideration in the next column, the addition of this column being performed usually by another full adder.

Problem 3 Describe the main features of a decade counter. A cascaded decade counter is to count up to 999 (decimal). Predict the binary number held in each block of the counter if the decimal count reads 374. Illustrate your answer with a diagram of the cascaded counter.

A decade counter comprises a set of four bistables and some logic gates which ensure a reset to zero of all bistables upon receipt of the tenth input pulse. A binary to decimal decoder is required if the count is to be displayed in decimal form, since, within the counter, the count is stored in binary form.

A cascaded counter to count up to 999 is illustrated in *Fig 10*.

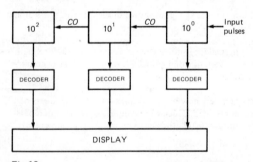

Fig 10

For a count of 374, the decades contain the binary equivalent of the components of the number. Thus:

the 10^0 decade contains 100 (binary equivalent of 4),

the 10^1 decade contains 111 (binary equivalent of 7), and

the 10^2 decade contains 011 (binary equivalent of 3).

C. FURTHER PROBLEMS ON BASIC DIGITAL CIRCUITS

(a) SHORT ANSWER PROBLEMS

1   In electronic circuits, logical 1 and 0 are represented by . . . . . .
2   A . . . . . . gate is one whose output is a logical 1 only if all its inputs are logical 0.
3   A NOT gate converts one . . . . . . state into the opposite state.
4   An OR gate can be constructed from a . . . . . . gate and a . . . . . . gate in series.
5   A half adder produces the sum and carry for . . . . . . binary digits.
6   A full adder produces the sum and carry for . . . . . . binary digits.
7   A truth table shows the logical . . . . . . state of a system for all possible . . . . . . of input states.
8   A decade counter is one which resets to . . . . . . upon receiving the . . . . . . input pulse.
9   A cascaded decade counter with 5 decades can count up to . . . . . . input pulses.
10 In order to read a cascaded decade counter, a . . . . . . and a . . . . . . are required.

(b) MULTI-CHOICE PROBLEMS (answers on page 126)

1   A two-input NAND gate produces an output:
    (a) $A + B$, (b) $A . B$, (c) $\overline{A + B}$, (d) $A + B$.
2   A NOT gate will:
    (a) not give an output, (b) not invert an input state, (c) not accept an input,
    (d) invert an input state.
3   Positive logic is characterised by:
    (a) quick results, (b) all outputs are positive, (c) positive potential representing
    logical 0, (d) positive potential representing logical 1.
4   A half adder will:
    (a) add two binary digits and produce a sum and carry, (b) add two binary digits
    and halve the result, (c) add half the sum to half the carry, (d) work at half speed.
5   A full adder will:
    (a) add two full binary numbers simultaneously, (b) add two binary digits and a
    carry from a previous addition, (c) recirculate the carry, (d) always have a logical
    1 carry output.
6   A decade counter will:
    (a) count to 9 and reset on the next pulse, (b) revert on the pulse after the tenth,
    (c) count for ten years, (d) count ten times faster than a binary counter.
7   A decade counter after 6 input pulses has its bistable set as follows (2^0 bistable
    on right):
    (a) 0 1 0 1, (b) 1 0 0 1, (c) 0 1 1 1, (d) 0 1 1 0.
8   A cascaded decade counter with x blocks will count up to:
    (a) $10x$ pulses, (b) $10-x$ pulses, (c) $10^x - 1$ pulses, (d) $10^{(x-1)}$ pulses

(c) CONVENTIONAL PROBLEMS

1   Using NOR, NAND and NOT gates, devise a system to produce the two outputs
    $(A . B) . (\overline{A + B})$ and $(A + B) . (\overline{A + B})$ with inputs A and B.
2   Use a truth table to verify that the sum S and carry C of two binary digits A and
    B are:
    $$S = A . \overline{B} + B . \overline{A}$$
    and $C = A . B$
    Devise a circuit using AND, OR and NOT gates to produce these results.

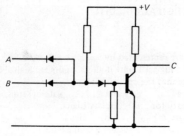

Fig 11

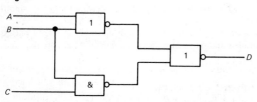

Fig 12

3 *Fig 11* shows a circuit for a simple NAND gate. Describe the circuit operation.

4 By inspecting the circuit given in *Fig 12*, deduce the expected output.

$$[D = \overline{(A + B) + \overline{B} \cdot C}]$$

5 Sketch a cascaded decade counter capable of counting to 9999. Predict the binary states of each block if the number of input pulses is 8 3 1 9.

[1000 011 0001 1001]

10 Feedback systems

A. MAIN POINTS CONCERNED WITH FEEDBACK SYSTEMS

1   **Feedback** in a system consists of connecting together two points in the system so
    that the signal appearing at a later stage is used to provide an input to an earlier
    stage. *Fig 1* illustrates the principle of feedback.
    The signal appearing at the output of stage 4 is fed back to the input of stage 2.

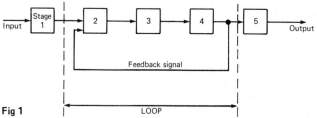

Fig 1

2   A **feedback loop** contains the entire signal path of that part of a system over which
    feedback occurs. This is also illustrated in *Fig 1*.
3   **Forward gain** is the ratio of output quantity/input quantity with all feedback loops
    broken. This has both amplitude and phase and is therefore generally a function of
    frequency. The symbol is G (see chapter 8).
4   **Closed loop gain** is the ratio of output quantity to input quantity with all feedback
    loops intact. For a system with a single overall feedback loop the loop gain is

$$G_C = \frac{G}{1 - \beta G}$$

where β is the fraction of the output which is fed back to the input. *Fig 2*
illustrates closed loop gain.
The total input to the system is the sum of θ_1 and $\beta\theta_0$.
This is amplified by G and appears at the output as θ_0.

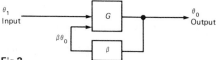

Fig 2

Thus $\theta_0 = G[\theta_1 + \beta\theta_0]$

i.e. $\theta_0[1 - \beta G] = G \cdot \theta_1$

$$\frac{\theta_0}{\theta_1} = G_C = \frac{G}{1 - \beta G}$$

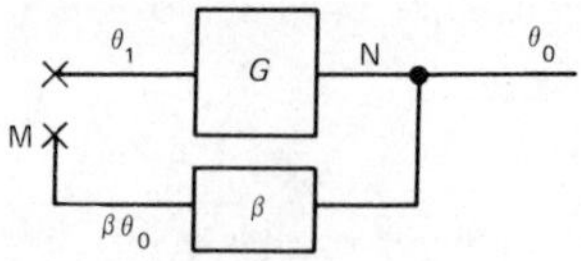

Fig 3

5 **Loop Gain** G_L is total gain experienced by a signal which is passed once round a loop. *Fig 3* illustrates loop gain.

Loop gain is also known as open loop gain.

θ_1 is amplified by G to produce $\theta_1 \cdot G$ at point N in *Fig 3* and attenuated (i.e. reduced) by β to produce $\theta_1 \cdot G \cdot \beta$ at point M in *Fig 3*. The total gain is the ratio of the signal at M to the input signal,

i.e. $\dfrac{\beta\theta_0}{\theta_1} = \dfrac{\beta G\theta_1}{\theta_1} = \beta G$

Thus, $G_L = \beta G$

Both β and G can involve phase change due to frequency change. However, in most elementary systems β does not change its phase angle and the remainder of this chapter assumes β to be a pure number, involving no phase change. This value of β is numerically less than one, since it represents the fraction of output fed back to the input, i.e., $\beta < 1$

6 **Negative feedback** is the condition where the signal fed back ($\beta\theta_0$ in *Fig 2*) is in antiphase, i.e., $180°$ out of phase with the input (θ_1). Generally, in a system there is a range of frequencies over which this condition can be met.

As the closed loop gain $G_C = \dfrac{G}{1 - \beta G}$ within this range, (see para 4), and $\beta G \theta_0$ is in antiphase with θ_1, then either G is negative with β positive, or β is negative with G positive. For negative feedback, the loop gain G_L is negative. If β is negative then

$G_C = \dfrac{G}{1 + \beta G}$, i.e., G_C is positive,

indicating that θ_0 remains in phase with θ_1.

For G negative: $G = \dfrac{-G}{1 + \beta G}$ and G_L is negative.

This implies θ_0 is in antiphase with θ_1. Note that G_C has the same value in both cases, but is either a positive or a negative number.

From the above, for the negative feedback case $|G_C| < |G|$, i.e. the system gain is **reduced** by the fraction $|1 + \beta G|$.

Note: $|x|$ means the positive value of x.

7 **Positive feedback** is the condition where the signal fed back is in phase with the input (θ_1).

Thus since $G_C = \dfrac{G}{1 - \beta G}$, and if $\beta G \theta_0$ is in phase with θ_1, then either β and G are both positive or β and G are both negative. The loop gain G_L is positive.

For β and G both positive, $G_C = \dfrac{G}{1 - \beta G}$, which implies θ_0 is in phase with θ_1.

For β and G both negative, $G_C = \dfrac{-G}{1 - (-\beta)(-G)} = \dfrac{-G}{1 - \beta G}$

which implies θ_0 is in antiphase with θ_1.

In either case the magnitude of C_C is the same, but is either a positive or a negative number.

Thus for the positive feedback case $|G_C| > |G|$, the system gain is increased by the fraction $|1 - \beta G|$.

112

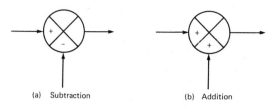

(a) Subtraction   (b) Addition

Fig 4

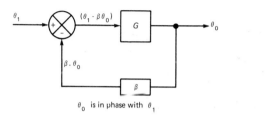

θ_0 is in phase with θ_1

Fig 5

8　A summing junction is the point in a system where input and feedback signals are combined. In block diagrams it is convenient to separate this from other blocks although in real systems it is not always easy to identify separate components which perform the addition. *Fig 4* shows the symbol for a summing junction and *Fig 5* illustrates a negative feedback system in block diagram form using this symbol.

9　A parameter is oscillating if it is varying in a regular periodic manner with time. Oscillating parameters can be represented graphically as, for example, in *Fig 6*. Many other oscillation waveforms exist.

10　A system is said to be oscillating continuously if its output is oscillating with constant amplitude and with no externally supplied input signal.

11　The conditions for a system to oscillate are:
(a) There must be feedback, (b) there must be a frequency or a range of frequencies over which the feedback is positive, (c) the loop gain (see para. 5) must be equal to 1.0.

In *Fig 7* the feedback is positive and it provides the only signal to the system input. For a positive feedback system, $G_C = \dfrac{G}{1 - \beta G}$. (See para. 7)

If $\beta G = 1.0$, then $G_C = \dfrac{G}{1 - 1} \to \infty$

For a system having infinite gain an output can exist for zero input, which permits θ_0 to take values other than zero.

Amplitude of Oscillation In a practical system, G_C can be very large while θ_0 is still fairly close to zero (or its 'rest' value). A small amount of disturbance at the input (e.g. thermal noise in an electronic amplifier or mechanical vibration in a position control system) results in a large movement of θ_0. As θ_0 increases however, the practical limits of output swing are reached (e.g. the limiting d.c. supply voltage of an amplifier). As θ_0 approaches this limit the forward system 'saturates' and the value of G falls rapidly until the loop gain is small. Since βG

113

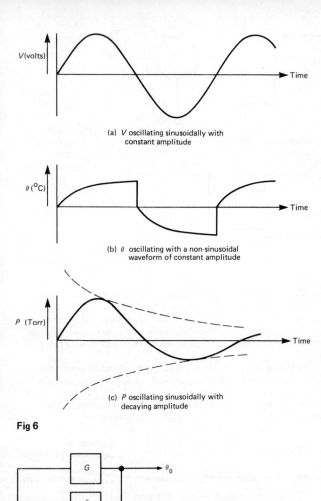

(a) V oscillating sinusoidally with constant amplitude

(b) θ oscillating with a non-sinusoidal waveform of constant amplitude

(c) P oscillating sinusoidally with decaying amplitude

Fig 6

Fig 7

is now small, the output begins to move rapidly back towards zero, this movement being enhanced by βG becoming again equal to 1.0 and the feedback being positive. θ_0 then moves rapidly to the other extreme and the process repeats. *Fig 8* illustrates this process.

The timing of the waveform depends upon the internal components in the system, and in some systems the output simply runs into saturation (positive or negative) and stays there. This condition is known as 'latch-up' and will not be dealt with here.

Fig 8

Sinusoidal Oscillation. The condition $\beta G = 1$ is met at only one frequency, which is the frequency of the sinusoidal oscillations. This means that there will be a frequency discriminating section in either the forward path or the feedback path.

12 **System Instability.** Whereas many systems are designed to oscillate to produce a variety of output waveforms, a great number of feedback systems are designed to respond to a deliberate input signal. Sometimes such a system oscillates accidentally, and it is then said to be **unstable**.

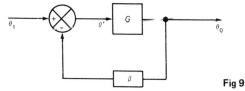

Fig 9

Fig 9 shows a system designed for negative feedback. θ_0 is in phase with θ' and β provides no phase shift. The summing junction subtracts $\beta\theta_0$ from θ_1, giving $\theta' = \theta_1 - \beta\theta_0$.

Since G is a function of frequency, its magnitude and phase angle vary with frequency. There will therefore be a frequency ω_0 at which θ_0 is $180°$ out of phase with θ'.

Under these circumstances θ_0 is negative with respect to θ', and $\theta' = \theta_1 - \beta(-\theta_0) = \theta_1 + \beta\theta_0$.

The feedback is thus positive at frequency ω_0. Under these conditions, if the loop gain is 1.0 the system is unstable.

13 **Mechanical Resonance.** A mechanical system comprising mass and a spring will oscillate if disturbed from rest. If the disturbance is of the right frequency then large amplitudes of oscillation will occur. This condition is called **resonance**. examples:

   (a) Out of balance road wheels on a car. At the right speed the out-of-balance vibration is amplified by resonance of the wheel mass and suspension and large movements of the wheel occur.

   (b) An electrically sustained tuning fork will resonate if the frequency of the electricity supply is correct for the mass and spring stiffness of the fork.

B. WORKED PROBLEMS ON FEEDBACK SYSTEMS

Problem 1 Define closed loop gain of a feedback system. If a feedback system has a forward gain of 40 and a feedback fraction of 0.017, calculate the closed loop gain if the feedback is (a) positive and (b) negative.

Closed loop gain of a feedback system is defined as the ratio of output quantity to input quantity with all feedback loops intact. It incorporates the phase angles of output and feedback signals.

(a) With reference to para. 7, positive feedback is given by

$$|G_C| = \left|\frac{G}{1 - \beta G}\right| = \frac{40}{1 - (0.017 \times 40)} = \frac{40}{0.32} = 125$$

(b) With reference to para. 6, negative feedback is given by

$$|G_C| = \left|\frac{G}{1 + \beta G}\right| = \frac{40}{1 + (0.017 \times 40)} = \frac{40}{1.68} = 23.8$$

Problem 2 The amplifier shown in *Fig 10* has a feedback signal derived from the two resistors R_1 and R_2. Determine the feedback fraction formula and the value of β if $R_1 = 100 \text{ k}\Omega$ and $R_2 = 2.7 \text{ k}\Omega$.

R_1 and R_2 divide the voltage V_0 in the ratio of R_2 to the total resistance,

thus $\beta V_0 = \left(\dfrac{R_2}{R_1 + R_2}\right) V_0$

$\beta = \dfrac{R_2}{R_2 + R_1}$

Inserting values, $\beta = \dfrac{2.7 \times 10^3}{(2.7 + 100) \times 10^3}$

$= 0.0263$ or 2.63%

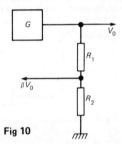

Fig 10

Problem 3 Shows that for a negative feedback system with a large value of loop gain the closed loop gain is practically independent of gain of the forward path.

With reference to para. 6, for a negative feedback system,

$$G_C = \frac{G}{1 + \beta G}$$

If $\beta G \gg 1$ then the denominator term becomes very near to βG.

Then $G_C = \dfrac{G}{\beta G} = \dfrac{1}{\beta}$

Thus G_C becomes independent of G for large values of βG.

Note: Since β (the fraction of the output signal which is fed back) is less than 1, the above result implies that the forward path gain G is very large. For practical purposes this result holds well for $\beta G > 20$.

C. FURTHER PROBLEMS ON FEEDBACK SYSTEMS

(a) SHORT ANSWER PROBLEMS

1 In a feedback system a . . . . . . of the output signal is fed back and . . . . . . to the input signal.

2 A feedback loop is comprised of the entire . . . . . . of a system in which . . . . . . occurs.

116

3 Forward gain is the ratio of . . . . . . to . . . . . . with feedback loops . . . . . .
4 Closed loop gain is the ratio of . . . . . . to . . . . . . with feedback loops . . . . . .
5 The distinction between forward gain and loop gain is that forward gain refers to signal amplification in the . . . . . . path and loop gain includes signal amplification in the . . . . . . path and signal attentuation in the . . . . . . path.
6 Forward gain, loop gain and closed loop gain are all functions of . . . . . .
7 For most practical systems there exists a range of . . . . . . for which gain and phase are constant.
8 In a negative feedback system the closed loop gain is . . . . . . and in a positive feedback system it is . . . . . . when compared with forward gain.
9 A feedback system will oscillate at a single frequency if the . . . . . . is unity and there is . . . . . . section in the system.

(b) MULTI-CHOICE PROBLEMS (answers on page 126)

In problems 1 to 9, select the correct answer from those suggested.
1 The output signal in a feedback system is:
(a) amplified, (b) connected to the input via an attenuator, (c) used in place of an input, (d) divided by the number of stages
2 The feedback network of a system:
(a) connects a fraction of the output to the input, (b) connects a fraction of the output to a summing junction, (c) adds the whole of the output to the input, (d) attentuates the input.
3 A positive feedback system:
(a) always has its output in phase with its input, (b) always has its output in antiphase with its input, (c) always has a positive output, (d) always has a positive loop gain.
4 A negative feedback system:
(a) has a large closed loop gain, (b) has a large forward gain, (c) has a gain less than the forward gain, (d) has a gain greater than the forward gain.
5 A negative feedback system having large loop gain:
(a) has a closed loop gain independent of forward path gain, (b) has a closed loop gain of unity, (c) has a closed loop gain independent of feedback fraction, (d) has a feedback fraction of unity.
6 The forward gain of a system with positive feedback is:
(a) divided by the amount $|1 - \beta G|$, (b) multiplied by the amount $|1 + \beta G|$, (c) divided by the amount $|1 + \beta G|$, (d) multiplied by the amount $|1 - \beta G|$.
7 A summing junction:
(a) always adds two inputs to give an output, (b) always subtracts one input from the other, (c) inverts the phase of one input with respect to the other, (d) provides an output which can be either the sum or the difference of two signals.
8 In order to oscillate, a feedback system must have:
(a) zero loop gain, (b) infinite loop gain with negative feedback, (c) unity loop gain, (d) unity loop gain with positive feedback.
9 The amplitude of oscillation in a system is determined by:
(a) the amount of feedback, (b) the physical limits of the system, (c) the open loop gain, (d) the frequency of oscillation.

(c) CONVENTIONAL PROBLEMS

1 Show that for a feedback system having forward gain G and feedback fraction β,

the closed loop gain is greater than the forward gain if the loop gain function is positive.

2   A system consists of two stages in the forward path having gains of 20 and 12. Determine the closed loop gain if 2% negative feedback is applied.

[41.4]

3   A negative feedback amplifier has a forward gain of 10^3 and a feedback of 4% of the output. Calculate the percentage increase in forward gain to achieve a 1% increase in closed loop gain. Comment on this result.

[68%]

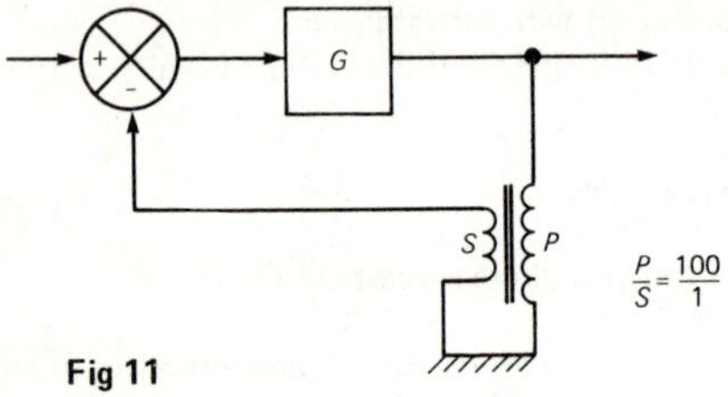

Fig 11

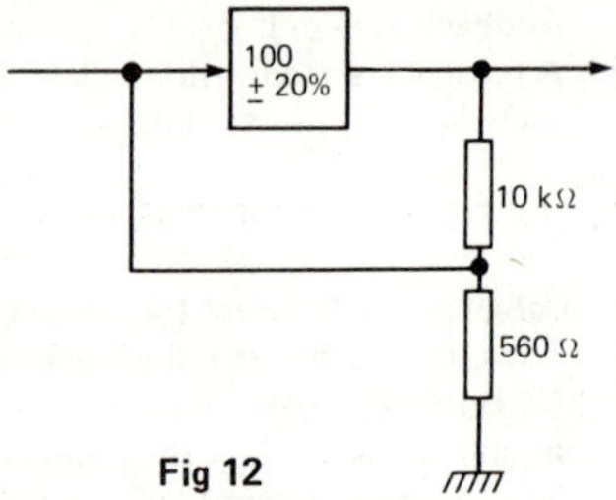

$$\frac{P}{S} = \frac{100}{1}$$

Fig 12

4   The circuit in *Fig 11* has a closed loop gain of 25. The turns ratio on the transformer is 100:1 and the system is connected for positive feedback. Determine the forward gain.

[33.3]

5   A batch of amplifiers to the same design has a ± 20% tolerance on a forward gain of 100. The negative feedback loop is closed by a pair of close-tolerance resistors as shown in *Fig 12*. Determine the overall closed loop gain tolerances for the batch.

$$\begin{bmatrix} + 2.7\% \\ -3.8\% \end{bmatrix}$$

6   An amplifier with a very large gain is found to distort a signal so that its output is not a good replica of its input signal. When negative feedback is applied such that the loop gain is much larger than 1.0 by using a resistor network as shown in *Fig 12*, the output is found to be a near perfect replica of the input. Explain this result.

7   State the condition for a feedback system to oscillate. Describe what steps need to be taken to obtain sinusoidal oscillations. Using this, explain why an out-of-balance road wheel of a car will oscillate at a certain road speed with a very good approximation to a sine wave.

11 Practical feedback systems

A. MAIN POINTS CONCERNED WITH PRACTICAL FEEDBACK SYSTEMS

1   A **sensor** is a device which detects the condition or change of condition of a parameter, such as temperature or voltage.

2   A **transducer** is a device which converts one form of energy into another. For example, a potentiometer supplied by a reference voltage converts an angular shaft position into an electrical potential.

3   A **reference input** is an external signal applied to the system used to obtain a specified response from the system.

4   A **command** is an input signal which may not have the same energy form as the feedback signal. The command then requires a transducer to convert it into its equivalent reference signal.

At the summing junction shown in *Fig 1* the two signals which form the forward path input must have the same energy form.

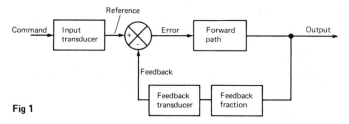

Fig 1

5   The **error signal** is the difference between the reference signal and the feedback signal. It represents directly the difference between the command and the output.

6   A **thermostat** is a temperature sensitive switch. It is a transducer in which electrical contacts are closed when the temperature of the sensing element falls below a temperature T_1 and are opened when the temperature of the sensing element rises above a temperature T_2, where $T_2 > T_1$.

7   A **tachogenerator** is a transducer which produces an electrical potential output which is directly proportional to the angular speed of rotation of a shaft.

B. WORKED PROBLEMS ON PRACTICAL FEEDBACK SYSTEMS

Problem 1 In terms of a feedback system, describe thermostatic control of the temperature of an enclosure.

The temperature of an enclosed space can be controlled using the following elements:

A heat source (e.g. a gas boiler), a fuel controller (e.g. an electrical solenoid valve), a thermostat, and a source of electrical energy.

Fig 2 shows the schematic control system.

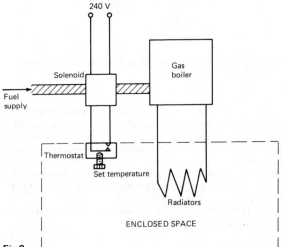

Fig 2

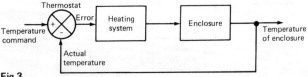

Fig 3

The system can be shown in block diagram form as in *Fig 3*.

The thermostat compares the actual enclosure temperature with the desired temperature and switches the heating on or off.

This type of control is known as ON-OFF control. No attempt is made to regulate the heat output. It is either fully on or completely off.

Problem 2 Describe a simple angular velocity control system using a feedback system.

A simple angular velocity control system is shown in *Fig 4*.

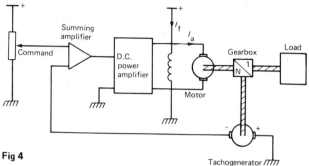

Fig 4

Tachogenerator

The command to change speed is made by turning a pointer over a scale of a potentiometer to the setting required. The potentiometer converts the setting into a d.c. voltage proportional to the commanded speed. Assuming the system starts from rest, the tachogenerator is not generating and there is zero feedback. The summing amplifier feeds a large error signal to the power amplifier, which in turn drives a large current into the armature of the d.c. motor. The motor field is separately excited with a constant field current I_f. As the motor speed increases the tachogenerator output rises, providing an increasing negative feedback input to the summing amplifier. When the feedback signal and the command signal are balanced there is no net input to the summing amplifier and the input to the power amplifier becomes zero. This represents the 'zero error' condition at which the motor is turning at a desired speed. Any speed reduction, perhaps due to changing load, results in an error signal which drives the power amplifier such that the error is removed. In practice, a small residual error is inevitable since some armature current is required to overcome friction and/or linkage losses. The motor therefore runs slightly more slowly than the desired value.

Problem 3 A linear potentiometer is to be used as a transducer, converting angular position to a d.c. potential. Sketch a suitable arrangement and deduce the law relating input angle and output potential. (Assume the potentiometer is not loaded.)
If the potentiometer has a 270° track and the input angle is 45°, what output voltage will be obtained if the d.c. supply is +5 V?

With reference to *Fig 5*, since the potentiometer is linear,
$V_0 \propto \theta_1$
If the maximum shaft angle is θ_m, then
$$\frac{V_0}{V} = \frac{\theta_1}{\theta_m}$$
$$V_0 = \left(\frac{\theta_1}{\theta_m}\right) V$$

Inserting values, $V_0 = \dfrac{45}{270} \times 5$

$V_0 = 0.83$ volts

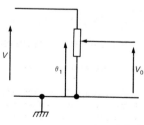

Fig 5

θ_1 = input shaft angle
V_0 = output voltage

121

Problem 4 Sketch a diagram of an angular position control system which uses potentiometers as input and feedback transducers. Explain briefly the operation of the system.

A sketch of the angular position control system is shown in *Fig 6*. Assuming that the angular shaft positions of the two potentiometers are identical, then the potential at the input potentiometer slider is equal in magnitude and opposite in sign to the potential at the feedback potentiometer slider.

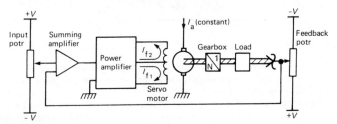

Fig 6

Under these conditions the output from the summing amplifier is zero and the power amplifier supplies balanced field currents I_{f1} and I_{f2} (i.e. the net field is zero). The system remains at rest. If now the input potentiometer shaft is rotated, its potential is altered. Assuming the input potentiometer potential moves positively, then the summing amplifier drives the power amplifier which in turn unbalances I_{f1} and I_{f2}. The net field is now not zero. The motor rotates and moves the output shaft so that the feedback potentiometer slider becomes more negative. When the output has moved far enough, there is again a balance between the two potentiometers and the system comes to rest.

The system always responds so as to reduce the difference in angle between the two potentiometers (shafts). The motor must be connected so that it rotates in the correct direction in response to an error input. The d.c. supplies to the potentiometer must be stable and equally balanced about earth (i.e. the signal earth of the system).

Note: The armature of a d.c. servomotor is supplied with a constant current. Under this condition, the torque is directly proportional to the field strength, so that the two halves of the split field control the motor in both direction and magnitude of torque. When the field is balanced, the torque is zero.

C. FURTHER PROBLEMS ON PRACTICAL FEEDBACK SYSTEMS

(a) SHORT ANSWER PROBLEMS

1 A thermostatically controlled system is one example of an . . . . . . system.
2 A transducer converts from one . . . . . . to another.
3 A potentiometer transducer converts . . . . . into . . . . . .
4 In a closed loop system the input and feedback transducers convert the . . . . . . signal and the . . . . . . into the same . . . . form.

122

5   In the speed control system in *Fig 4* the input command is an . . . . . . of the input potentiometer and the output is the . . . . . . of the load.
6   A control system responds to an . . . . . . signal generated by the difference between . . . . and . . . .
7   A position control system is one in which the . . . . . . of the load is required to align with the . . . .
8   In a split-field servomotor with constant armature current, the output . . . . . . is proportional to the net . . . . . .

(b) MULTI-CHOICE PROBLEMS (answers on page 126)

1   In a control system the input is:
    (a) a command signal, (b) a reference signal, (c) an error signal, (d) a transducer output.
2   A control system operates:
    (a) to move the load, (b) to provide feedback, (c) to reduce error signals, (d) to test transducers.
3   In a control system an error signal:
    (a) is always avoided, (b) is desirable, (c) is the difference between command and output, (d) is the difference between reference and feedback signals.
4   The purpose of a power amplifier in a control system is to:
    (a) switch a motor on, (b) provide electrical power proportional to the error signal, (c) reverse the motor torque, (d) provide electrical power to the load.
5   An automatic control system is characterised by the following:
    (a) it is an error-actuated system, (b) its output is proportional to its input, (c) it contains transducers, (d) its output is larger than its input.

(c) CONVENTIONAL PROBLEMS

1   Sketch the main features of an electric toaster. By considering the block diagram of the toaster, determine the input and output quantities.
2   Devise a simple door-opening mechanism which operates on the interruption of a beam of light. Sketch the block diagram and identify input, forward path, output and feedback elements. Is this a continuous or an on-off system?
3   In driving a car, the following are the main features:

Command	=	road direction
Input transducer	=	eyes
Summing junction	=	brain
Power amplifier	=	muscles of arms and hands operating on steering wheel
Output	=	vehicle direction
Feedback transducer	=	eyes

Sketch and label a block diagram of the feedback system implied by the above features.
(Note: Biological feedback systems often use the same organs simultaneously to perform different functions. The eyes in this problem observe both the actual heading of the car and the immediate future heading of the road. The brain also performs more tasks than simply comparing the input and feedback information.)
4   Outline a system to provide constant lighting conditions in a room using a thyristor lamp dimmer. The dimmer supplies an electric light which ranges from fully off in bright daylight conditions to fully on at night. (Assume that the maximum lamp brightness produces equivalent lighting to bright daylight.)

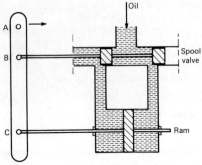

Fig 7

5 *Fig 7* shows a schematic arrangement of a spool valve and a hydraulic ram such as might be used in a power braking system of a car. Describe, by referring to the movements of points B and C in response to an input displacement of A, the operation of the system.

6 A governor mechanism is used to control fuel flow to an engine so as to regulate its speed. Sketch an outline operational diagram of the system and explain its operation. Suggest how the engine speed can be adjusted using the regulator.

Answers to multi-choice problems

CHAPTER 1 (*page 9*)

   1 (b);   2 (e);   3 (e);   4 (c);   5 (c);   6 (a);   7 (d);
   8 (f);   9 (b); 10 (c), 11 (b), 12 (a); 13 (b), 14 (h).

CHAPTER 2 (*page 22*)

   1 (c);   2 (d);   3 (d);   4 (a);   5 (d);   6 (c);   7 (b);
   8 (c),   9 (b); 10 (c).

CHAPTER 3 (*page 32*)

   1 (c);   2 (a);   3 (b);   4 (b);   5 (a);   6 (d);   7 (d);
   8 (b);   9 (c); 10 (b).

CHAPTER 4 (*page 45*)

   1 (d);   2 (c);   3 (d);   4 (b);   5 (b).

CHAPTER 5 (*page 64*)

   1 (d);   2 (g);   3 (i);   4 (s);   5 (h);   6 (b),   7 (k);
   8 (1);   9 (a); 10 (d). (g), (i) and (1) 11 (b); 12 (d).

CHAPTER 6 (*page 83*)

   1 (g);   2 (c);   3 (a),   4 (a);   5 (f);   6 (a);   7 (g);
   8 (1);   9 (1), 10 (d), 11 (f); 12 (j);

CHAPTER 7 (*page 91*)

   1 (b) and (c)   2 (g);   3 (d);   4 (a),   5 (h);   6 (k);
   7 (j)   8 (f);   9 (c).

CHAPTER 8 (*page 99*)
   1 (c);   2 (c);   3 (d),   4 (a);   5 (a);   6 (b),   7 (c);
   8 (a).

CHAPTER 9 (*page 109*)

    1 (b),   2 (d);   3 (d);   4 (a);   5 (b);   6 (a);   7 (d);
    8 (c).

CHAPTER 10 (*page 117*)

    1 (b);   2 (b);   3 (d);   4 (c);   5 (a);   6 (a);   7 (d);
    8 (d),   9 (b).

CHAPTER 11 (*page 123*)

    1 (a);   2 (c);   3 (d);   4 (b);   5 (a).

Index